SpringerBriefs in Ethics

Springer Briefs in Ethics envisions a series of short publications in areas such as business ethics, bioethics, science and engineering ethics, food and agricultural ethics, environmental ethics, human rights and the like. The intention is to present concise summaries of cutting-edge research and practical applications across a wide spectrum.

Springer Briefs in Ethics are seen as complementing monographs and journal articles with compact volumes of 50 to 125 pages, covering a wide range of content from professional to academic. Typical topics might include:

- Timely reports on state-of-the art analytical techniques
- A bridge between new research results, as published in journal articles, and a contextual literature review
- A snapshot of a hot or emerging topic
- In-depth case studies or clinical examples
- Presentations of core concepts that students must understand in order to make independent contributions

More information about this series at http://www.springer.com/series/10184

H. Russell Searight

Ethical Challenges in Multi-Cultural Patient Care

Cross Cultural Issues at the End of Life

 Springer

H. Russell Searight
Department of Psychology
Lake Superior State University
Sault Sainte Marie, MI, USA

ISSN 2211-8101   ISSN 2211-811X   (electronic)
SpringerBriefs in Ethics
ISBN 978-3-030-23543-7   ISBN 978-3-030-23544-4   (eBook)
https://doi.org/10.1007/978-3-030-23544-4

This Springer imprint is published by the registered company Springer Nature Switzerland AG
The registered company address is: Gewerbestrasse 11, 6330 Cham, Switzerland

Contents

Chapter 1
History and Background of End-of-Life Decision-Making and Culture

Don't talk that way.

This admonition from members of the Navajo community arose in response to the discussions of terminal illness and a patient's desire for life-support—both of which are implicit topics in discussions of advance directives. While required by federal law in the United States, discussions of advance directives were seen as harmful by Navajo elders:

.... You don't say those things. And you don't try to bestow that upon yourself, ... The object is to live as long as possible here on earth. Why try to shorten it by bestowing things upon yourself (Carrese & Rhodes, 1995; p. 828).

The extension of life through reduced mortality from infectious disease, and the increased sophisticated of medical technology have all contributed to the possibility that we will, in the future, be in a state with a less than optimal quality of life and questionable chance of recovery. While we may not be conscious, we will likely be maintained alive with the assistance of ventilators, tube feeding, artificial hydration, and other medical interventions. In the United States and most Western countries, the past 50 years have seen a major shift away from physician authority to patient centered decision-making. As a result, we are in an historical period in which we can extend our autonomy regarding medical care to situations in which we are physically no longer able to express our wishes. The discussion on the Navajo reservation noted above was prompted by federal guidelines based on the Patient Self-Determination Act (PSDA), that require healthcare facilities to raise the topic of advance care planning with patients and encourage patients to complete an advance directive.

The implementation of the PSDA in the United States in the 1990s highlighted some of the core issues and the diversity of perspectives in end-of-life decision-making. These issues include whether we would want aggressive care to treat conditions such as pancreatic cancer or instead, accept inevitable death and request palliative care focusing on comfort measures. Because of our increased longevity and the rise in neurocognitive disorders with age, many of us may not have the cognitive ability to make a medical decision when the time arises. Patients who are

© The Author(s), under exclusive license to Springer Nature Switzerland AG 2019
H. R. Searight, *Ethical Challenges in Multi-Cultural Patient Care*,
SpringerBriefs in Ethics, https://doi.org/10.1007/978-3-030-23544-4_1

comatose and on life support obviously do not have the ability to render a current decision about desired medical care.

Additionally, when the PSDA was implemented with some ethnic and cultural communities in the United States, such as the Navajo, it was not well received by patients or their families. The PSDA assumed that patients would want to make decisions about their own care. However in many Asian, Latin American, and southern European countries, the patient's family, rather than the patient, themselves, receives information about the patient's diagnosis and makes treatment decisions on the patient's behalf. It was also found that among other groups, such as African–Americans in the United States, the idea of medical treatment being "futile" and prolonging a loved one's suffering was not congruent with the spiritually-influenced worldview of the patient and their family.

1.1 Seminal Cases Highlighting the Importance of Advance Care Planning

Interest on the part of the professional health care community as well as the general public in preparing for a time of terminal illness in which one can no longer voice their wishes for various forms of life support as well as other potentially life-sustaining medical interventions, stems from several well-publicized legal cases in the United States. In these situations, relatively young patients were severely neurologically impaired and unable to communicate. While being maintained through artificial nutrition, hydration and ventilation, these patients demonstrated little likelihood of recovery.

1.1.1 Karen Ann Quinlan

In April, 1975, Karen Ann Quinlan, a 21-year-old woman, was admitted to the hospital unconscious after reportedly ingesting a combination of tranquilizers and alcohol. During the hospital course, she lost 50% of her body weight from 120 to 60 lb and was placed on a respirator and fed through a tube. In November of that year, Ms. Quinlan's parents sought to have the respirator removed. The informal practice of physicians discontinuing life-support for comatose patients was still relatively common (Pence, 2003, 2016). However, in Ms. Quinlan's case, her parents directly asked the physician caring for their daughter to do so—a practice that was unusual. In response, the doctors refused the parents' request indicating that they were concerned about the possibility of legal action including being sued for malpractice. The parents then initiated legal proceedings. After a State Superior Court judge denied the parents' request, that decision was reversed on appeal to the State Supreme Court. The State Supreme Court indicated that if there was no reasonable possibility that Ms. Quinlan would recover, the State's interest in maintaining her life was overridden by Ms. Quinlan's interest in not continuing to be maintained alive artificially in her current state (Pence, 2016). However the court-appointed representative for Ms. Quinlan argued that withdrawing life support resulting in her death would be homicide. Ms. Quinlan was receiving care in a facility affiliated with the Roman Catholic Church. The nuns overseeing Ms. Quinlan's care were also opposed to removing the ventilator. In his analysis, Pence (2016) argues that the local Church's and the Catholic hospital's position may have been, at least partially, motivated by a US Supreme Court ruling two years earlier, *Roe versus Wade*, which supported a woman's right to terminate a pregnancy. Catholic theologians testifying in the Quinlan case asserted that there was no right to die and were concerned about the "slippery slope" into euthanasia that the situation implied.

Despite the final court ruling, Ms. Quinlan's ventilator was not abruptly stopped. Instead, she was weaned from the ventilator over the span of many weeks and done so in a way that allowed her to breathe on her own. Since Ms. Quinlan was comatose, the court indicated that the patient's father, rather than the physicians involved, was the appropriate decision-maker on her behalf. One of the legal opinions from a judge in the case stated that maintaining Ms. Quinlan alive constituted cruel and unusual punishment. It was predicted that Ms. Quinlan would readily succumb after the respirator was removed. However she was able to breathe on her own and did continue to receive nutrition via a feeding tube. Ms. Quinlan died of pneumonia ten years later in a nursing home (McFadden, 1985).

1.1.2 Nancy Cruzan

In 1983, 25-year-old Nancy Cruzan lost control of her car, ran off the road and ended up laying facedown in a ditch filled with water. It was reported that at the scene of the accident, Ms. Cruzan had no detectable vital signs. She reportedly stopped breathing for 15 min immediately after the accident but was subsequently resuscitated. Three weeks later, it was determined that Ms. Cruzan was in a persistent vegetative state and could not swallow A feeding tube was implanted (Pence, 2016). After nearly five years had passed, Ms. Cruzan's parents requested that the feeding tube be removed. However, the hospital was unwilling to do so because they were concerned that they could be prosecuted for murder. Ms. Cruzan had demonstrated little responsiveness and because of the anoxia associated with the accident, it was likely that her neurological functioning was seriously and permanently impaired (Pence, 2003). A Missouri probate court ruled that the feeding tube could be removed and emphasized that adult patients had the right to refuse or discontinue life-support. However, the State of Missouri intervened and appealed to the State Supreme Court which reversed the lower court's ruling. The State Supreme Court based their decision on the absence of clear and convincing evidence about what Ms. Cruzan would have wanted in these circumstances (Pence, 2016). The State Supreme Court indicated that no one can refuse treatment for another person unless the patient had previously communicated a clear choice which was well-documented. The case eventually went to the U.S. Supreme Court which acknowledged that competent individuals do indeed have the right to refuse treatment. However, if a patient's decisional abilities are impaired, clear and convincing evidence of their wishes, such as through a written advance directive developed when the individual was competent, was required for discontinuing treatment. In Ms. Cruzan's case, this information was initially unavailable. However, according to her parents, Ms. Cruzan had expressed to coworkers the desire to never "live like a vegetable." Cruzan's physician referred to her current state as a "living hell" and also recommended removal of the feeding tube. The testimony of friends and coworkers, which was unavailable in previous court decisions, met the lower court standards of "clear and convincing evidence" (Pence, 2016) and led to a reversal of the State Supreme Court's earlier decision. Shortly thereafter, the feeding tube was removed. The Cruzan case became very politicized with right-to-life supporters attempting to storm the hospital unit and reattach the feeding tube. Ms. Cruzan died approximately 12 days after removal of the feeding tube.

1.2   Implications of the Quinlan and Cruzan Cases

The U.S. had witnessed two particularly challenging cases in which women were being kept alive even though they had been nonresponsive for many years. These cases became known internationally and provoked discussion and debate among Europeran ethicicts (Kennedy, 1976; Truog, 2008). The key reason for maintaining their lives was that these young women had not indicated a preference about the level of care they desired if they were in nonresponsive/comatose state.

The protracted attention given to the Cruzan case in the media conveyed that medical science could maintain life for years even though the life being preserved did not exhibit evidence of the neurocognitive properties of awareness, communication, and basic reasoning. The plight of the Cruzan family clearly illustrated how others would make the decision about one's continued life if the patient could not communicate their wishes. However, expressing one's wishes indirectly through a past conversation with family members or friends was not a firm foundation for health care providers to act or discontinue treatment. Additionally, these cases illustrated that resulting legal proceedings could require years to resolve with the patient existing in a prolonged persistent vegetative state (PVS). The Cruzan decision triggered informal conversations among some of the general public about whether one would want to be maintained alive if they were in a similar state.

It also made all concerned aware that they could be placed in a position of having to represent the interests of an ill family member who had become incompetent—a burden that a husband or wife did not want nor did the spouse want to place a loved one under this duress. Statements such as those made by Cruzan's physician about her experience in an enduring non-responsive state ("a living hell") led to conversations that probed unknowns about medical science. For example, did individuals in a long-term coma of many years ever regain consciousness? While recovery of awareness became less likely as the coma endured, a very few patients actually did "wake up" years later (Pence, 2003). Physicians and hospital administrators were concerned about the legal implications of withdrawing life support—particularly being sued for malpractice or even charged with homicide.

The publicity surrounding the Cruzan case appeared to accelerate the creation of Do Not Resuscitate (DNR) policies that hospitals had begun developing in the early to mid-1970s. Cardiopulmonary resuscitation (CPR) arose in the 1960s as a technique for patients suffering anesthesia-related cardiac arrest. The technique was featured on some popular medical television dramas—usually with great success (Diem, Lantos, & Tulsky, 1996). CPR soon came to be employed on a much larger scale with many patients experiencing cardiopulmonary arrest arising from diverse causes. While later augmented by the PSDA, the DNR discussion became a standard part of many hospital and nursing home admissions as well as a precursor to some surgical procedures. Again, by formalizing their wishes, patients and/or families were able to specify how aggressive their care would be—even though the general public's knowledge of the success rates of many interventions was inaccurate (Diem, Lantos, & Tulsky, 1996).

1.3 The Patient Self-Determination Act (PSDA)

The central problem in the Cruzan case—clearly establishing the extent of care that one would want on their behalf if they were unable to convey their wishes—could be, it was argued, resolved with a clearly documented record of the patient's preferences established well in advance of life threatening illness. The mechanism of this communication, an advance directive for treatment, would be a record that would direct physicians to provide the level of care desired when the patient could not express their wishes. One form of advance directive, the living will, at least in theory, bypassed family members and lets the previously competent patient direct their own care.

The federal Patient Self-Determination Act (PSDA), according to many commentators, was a direct outgrowth of the public reaction to the Cruzan case. The PSDA required that hospitals receiving federal Medicaid funding have policies for asking all patients if they had an advance directive. If patients did not have a directive, hospital staff were to educate patients about these legal documents and encourage them to develop a record of their health care wishes. In early research on advance directives, there seemed to be a strong consensus among adult patients across the age spectrum that the documents were desirable. Additionally, there was a fairly pronounced preference that aggressive life sustaining measures would not be desired if one was comatose with a chance of survival (57%); in a persistent vegetative state (85%); suffering from dementia (79%) and having dementia with a terminal illness, (87%) (Emanuel, Barry, Stoeckle, Ettelson, & Emanuel, 1991).

1.4   The Doctrine of Informed Consent

Legally, the patient's right to make healthcare decisions for themselves, including declining recommended treatment such as refusing amputation of a gangrenous leg, has its roots in a ruling by Judge Cardozo in Schloendorff v. New York Hospital (1914) Cardozo asserted that … "every human being of adult years and sound mind has a right to determine what shall be done with his own body." In U.S. health care law and clinical practice, there is a strong emphasis on individual autonomy. A competent adult individual has the right to choose their medical treatment, including the choice not to receive medical treatment and to terminate ongoing treatment an example of the latter is the finding that between 8 and 31% of patients on dialysis stop treatment on their own (Qazi, Che, & Zhu, 2018).

However, it eventually became apparent that in order to make genuinely autonomous medical decisions, patients needed to be provided with adequate information. The basic tenets of informed consent included disclosure of the patient's diagnosis, the impact of condition on the patient's daily living, available treatment options with accompanying risks and benefits and the prognosis with and without treatment (Searight & Barbarash, 1994). Making healthcare decisions for oneself is also predicated on the assumption that the patient exhibits decision-making capacity and voluntarily provides consent (Walter, 1997). In order to have intact decisional capacity, Grisso and Appelbaum (1998) indicated that patients must (1) Be able to communicate A clear and consistent choice: (2) Understand relevant information; (3) Appreciate their current health situation and both its short-term and long-term consequences; (4) Based upon their understanding of the condition and its prognosis, engage in an internally logical consistent reasoning process for selecting a decision appreciate the information.

Physicians have been found liable for performing procedures on patients who have not been provided with all relevant information with particular attention to risks and foreseeable benefits. While patient autonomy, predicated on having adequate comprehensible information, has been the basis of this litigation, the physician's defense has focused on beneficence—acting in the patient's best medical interests. Patient autonomy may conflict with physicians' judgment about what was or would be most beneficial to the patient. Legally, physicians have been found liable in which they proceeded with treatment that they thought was in the patient's best interests without receiving the patient's explicit advance consent. For example, in *Dries vs. Gregor*, based upon a finding of a growth on their right breast, the patient agreed to a biopsy of the breast to determine the possible presence of cancer. However, during the biopsy, a partial mastectomy was conducted and substantial amount of breast tissue removed; the growth was found to be benign. Based upon available information, the physician argued that he had acted in the best interest of the patient to prevent potential cancer from spreading and also to avoid having to perform a second procedure. However, the patient had consented to only a much more conservative intervention (Walter, 1997)—namely, a biopsy and had not been informed that a more invasive procedure, resulting in removal of significant beast tissue, would be performed.

While not consistently upheld in US courts, there are circumstances in which physicians have argued that obtaining informed consent would adversely impact the patient. In intentionally choosing to not inform a patient of medical "bad news," a strong case has to be made that the disclosure of the information could reasonably be expected to adversely impact the patient's condition (Walter, 1997). As will be discussed later, this form of beneficence has been the basis of cultural values that support non-disclosure.

With the publicity given to the Quinlan and Cruzan cases as well as the PSDA, the elements of informed consent became "front and center" in patient decision-making. A potential solution was to have individuals make decisions for themselves in anticipation of being in the state in which they could not express themselves or make decisions. A document such as a living will or an advance directive formally appointing a durable power of attorney would both "speak" on the patient's behalf. In essence, theses proxies extended patient autonomy to situations in which the patient could no longer indicate their wishes. As the scenario on the Navajo reservation makes clear, both advance directives and living wills ask us to imagine what our values would be if we were in a state in which our chances of survival were poor or at best, unknown.

In cultures in which language, thought and action are inextricably linked, end of life discussions are not hypothetical future possibilities but, instead, give terminal illness and death a reality. In Carrese and Rhodes' (1995) study, a number of their informants would not even discuss advance care planning because of this perceived danger. This same perspective, while not as explicitly stated as among the Navajo, has been found in other cultures such as Bosnian immigrants (Searight & Gafford, 2005).

However, it was not solely patients who did not welcome discussions of possible impending death. In the medical community a patient's death is still often seen as a failure on the part of physicians. Intensive care nurses have observed that end-of life treatment decisions are often based on the physician's rather than the patient's needs: "Too many [physicians] see death as a personal affront to their professional abilities and do not visualize the dying process as part of life itself. Thus, many patients suffer needlessly without adequate pain control and supportive care" (Beckstrand, Callister, & Kirchoff, 2006; p. 41). As someone who has spent a great deal of time training physicians, it was not unusual to see a resident physician in-training coming off their shift of overnight call in the hospital breathing a sigh of relief because a terminally ill patient had not died "on their watch."

1.5 The Blackhall et al. Study

Within several years of the PSDA's initiation, a provocative study on cross-cultural end-of-life views by Blackhall and colleagues (Blackhall, Murphy, Frank, Michel, & Azen, 1995) was published in the *Journal of the American Medical Association.* The paper received a good deal of public attention in news accounts as well as served as an impetus for future research on cross-cultural issues in end-of-life decision-making.

Blackhall et al. (1995) recruited samples of older adults of African–American, Korean–American, Mexican–American and European–American background. The participants were asked if they believed that patients should be informed about a serious cancer diagnosis and whether they should be informed directly of their prognosis. In their groundbreaking work, these investigators found that there were major differences between these ethnic groups regarding end of life preferences. African–Americans and European Americans strongly endorsed the principle that patients should be told their diagnosis and the majority of both groups believed that patients should be informed of their prognosis even if it was terminal. Among Korean Americans, fewer than half (45%) indicated that a cancer diagnosis should be disclosed to patients with 38% indicating that they should be informed of their prognosis. For the Mexican American sample, approximately 60% indicated that the patient should be informed of the diagnosis and about 45% indicated that patients should be told of a terminal prognosis. When asked who should make decisions about whether to put the patient on life support, about 60% of African Americans indicated that the patient should do so while approximately 15% said the physician and 25% said the family should make this decision. For European–Americans, slightly more than 60% indicated that patients should be the primary decision-maker regarding life support with fewer than 10% viewing the physician as the source of these decisions and approximately 20% indicating that the family should be the primary decision-maker (Blackhall et al., 1995). In contrast, nearly 60% of the sample's Korean–Americans indicated that the family should make the decision about life support followed by close to equal representation of physician and patient as decision-makers. For Mexican Americans, about 50% indicated that the patient should make life support decisions while 10% indicated that the physician should decide with 45% indicating that these decisions were the family's role. Among the Mexican–American sample, those with fewer years of formal education and those who were older were most likely to indicate that the patient should not be told the truth nor should patients be the primary decision-makers about treatment for a life-threatening condition (Blackhall et al., 1995). There were suggestions that as Mexican Americans became more acculturated, they were more likely to endorse patient-centered decision-making (Blackhall et al., 1995).

1.6   Cross-Cultural Differences in Values at the End of Life

While these data were provocative, the survey format used by Blackhall et al. (1995) provided little information about the reasons for the diversity in values among ethnic groups. Follow up qualitative interview studies provided some insight into these differences. In an interview with an older Korean American woman, Mrs. Kim, it became evident that one's duty to family members was far more important than individual autonomy. In fact, it was incumbent upon the extended family to do anything to prevent their loved one's death (Frank et al., 1998).

Culturally based objections to ADs were sometimes less tangible—there was a feeling that one should not be forecasting the circumstances of ones' own death. As a recent Bosnian immigrant to the U.S. put it: "It's like playing with your destiny." (Searight & Gafford, 2005). Additionally, even hypothetical discussions about future death were seen as undesirable as illustrated by this exchange with Mrs. Kim:

Interviewer: "In order to use [a written advance directive]…, It is necessary to talk about death when the potential patient is still conscious and healthy what do you think about this?"
Mrs. Kim: "It is not good to talk about death in advance."
Interviewer: "Even for the older people?"
Mrs. Kim: "I am also old. Although the person may be old, to talk on the issue in advance is not good. If asked to sign such papers, without knowing one's future, how could I sign them?" (Frank et al., 1998; p. 423)

1.7 The Era of the Patient as an Autonomous Consumer

The contractual nature of advance directives may be seen as part of broader changes in the physician-patient relationship, as well as in the nature of medical care. These forces have likely contributed to a greater emphasis on patient autonomy. Changes in medicine include replacing the long-term family doctor-patient relationship with a more mobile physician work force. The family doctor depicted in Norman Rockwell paintings, who knows the patient and their family well, has been replaced by physicians who are employees of large healthcare organizations. As executive employees, contemporary physicians, compared with colleagues a generation earlier, more readily move from practice to practice within and between geographic regions. The rising number of urgent care centers and walk in clinics—often housed in strip malls next to retail stores or physically part of, chain-store pharmacies, while providing convenience, does little to foster a stable, familiar, medical "home." These clinics are based upon values of availability and efficiency rather than having a long-term relationship with the healthcare provider. Without this history of an enduring physician–patient relationship, trust in a physician's advice is likely to be received with greater tentativeness (Schlesinger, 2002). As public confidence in physician authority declined (Schlesinger, 2002; Tomes, 2016)—a pattern also seen in universal access countries such as Canada and Great Britain—patients could no longer be sure that their physician was looking out for them. As a result, patients needed to look out for themselves. Under these circumstances, it is understandable that patients may wish to take advance control over late life medical care–particularly when physically and emotionally vulnerable and unable to verbalize their preferences.

The consumer orientation and patient rights movement as well as the availability of specialized medical knowledge to the general public through the internet and pharmaceutical companies' direct marketing efforts (Schlesinger, 2002), are additional factors which have moved medical decision-making further into the patient's domain.

The growing emphasis on "patient as expert" contributed to a physician-patient relationship that is more collaborative than authority based. In the contemporary models of patient-centered care now widely taught in U.S. medical schools, physician advice is being replaced with shared physician-patient decision-making (Constand, Mac-Dermid, Dal Bello-Haas, & Law, 2014).

1.8   Medical Technology and the Rise of Subspecialty Care

With the shift from acute to chronic illness, it has been argued that the conditions that eventually lead to death—cancer and cardiovascular disease, in particular—are "natural," yet, prolonged, processes of ending lives no longer taken by rapid-onset infectious illness. Historically, early kidney transplants and the first successful heart transplant by Christian Barnard, underscored the ability to delay natural physical deterioration. Our ability to be maintained alive as we age and amidst overall deterioration in our general health status, has grown exponentially (Kaufman, 2015). While organ transplantation and mechanical support for physiological functions were not widely available in the past, these once-experimental treatments have become the standard of U.S. medical care. When kidney dialysis was introduced in the United States in Washington, a group of citizens, that came to be known as the "God Committee," had to decide on the basis of personal characteristics who should receive treatment and who would be denied dialysis (Markel, 2017). When the workings of the "God Committee" became public through a widely circulated news magazine, the response triggered rapid passage of federal legislation insuring kidney dialysis for all who needed it.

With the passage of Medicare—essentially a form of universal health coverage for senior citizens—in the mid 1960s, decisions needed to be made about which treatments would be covered. The rise of the evidence-based medicine movement in which research findings from clinical trials justify performing specific procedures as well as having them funded by third parties, also contributed to relying upon a growing array of therapies that could maintain life (Kaufman, 2015).

In many instances, it became unclear where to draw the line. Kaufman (2015) presents a case of Mrs. Dang, a 72-year-old Asian American woman with chronic liver disease. A liver specialist recommended that Ms. Dang receive a liver transplant. The physician indicated that while the first year post transplant would be difficult with the new liver, Mrs. Dang could live 10–15 years longer with no problems. One of Ms. Dang's daughters said "I need to ask my mother if she wants to live ten more years." (Kaufman, 2015; p. 41). Complicating the decision further for patients is that about one out of three liver transplant patients experience significant complications. Kaufman (2015), however, suggests that this aspect is downplayed when patients are presented with the option of a transplant. During the family decision-making process Ms. Dang's daughters raised an ethical question: "… if you have cancer and decide not to treat it, is that suicide? I don't think so, but I wonder. If I think my mother shouldn't be listed for the transplant, is that murder?" (Kaufman, 2015; p. 42). As

Kaufman (2015) points out, this type of dilemma only arose after Medicare began paying for liver transplants.

Simultaneous with the development of new technology came new subspecialties within medicine. Residency is no longer the end of formal medical education. Many physicians, after a 3 to 4-year residency within a medical specialty such as internal medicine or radiology, embark upon post-residency fellowship training focusing on subspecialty areas within medicine. Subspecialty training in medicine has become much more common. For example, a recent study found that approximately 90% of residents graduating from an orthopedic residency went on to fellowship training (Yin, Gandhi, Limpisvasti, Mohr, & ElAttrache, 2015). While this level of sophistication can no doubt benefit patients, it also leads to a fragmented, rather than comprehensive and holistic approach to patient care. When livers, hearts, knees, etc. are viewed in isolation, the patient, situated in a life context, can be neglected. For example, a urologist might propose a laser treatment for a narrowed urethra in an elderly patient with rapidly advancing mid-stage dementia or late stage metastatic cancer. While the laser treatment may indeed be the evidence-based treatment of choice for the condition, the fact that a confused older patient with a predicted limited lifespan will have to undergo treatments that requires them to lie perfectly still, may simply not be feasible in the context of the patient's cognitive impairment. Furthermore, when the patient's prognosis suggests a limited lifespan, it may be difficult to justify the additional psychosocial strain of the procedure as being in the patient's overall best interest. Yet, when the patient's urethra is viewed in isolation, the prognosis with treatment for the specific urological condition is quite positive.

Evidence also supports the limitations of a narrow view of organ specific treatment. Data suggest that among older patients with multiple medical conditions including kidney failure, dialysis did not improve the patient's likelihood of survival and also was associated with declining functioning among those in nursing home settings. However, as is the case with cancer patients, most dialysis patients receive aggressive, hospital-based treatment during the last month of life and die in the hospital (Combs et al. 2015).

While the broader social implications, including the patient's quality of life, are often neglected, the average age of those undergoing surgery and similarly invasive interventions in the United States has been increasing. For example, stents to open cardiac arteries are now routinely inserted among patients in their 80s and 90s. Kaufman (2015) notes that as more older patients undergo the procedure and with Medicare authorizing coverage for stent placement, not having a stent placed would become sub-standard care. At the same time, there is controversy about the effectiveness of stents with some medical investigators noting that they are no more effective in preventing cardiovascular deaths than medication or changes in diet and exercise (Kaufman, 2015; Redberg, Katz, & Grady, 2011).

In part, the rising influence of evidence-based medicine has also indirectly become a factor in prolonging life. When older patients with multiple medical problems receive a recommendation for kidney dialysis or transplant or for the installation of a pacemaker, these endorsements are based on research evidence. However, it is important to recognize that the samples in which these interventions were studied and determined to be effective, often do not reflect clinical reality. Typically, patients in medical research trials do not have a significant number of possible confounding conditions. However, the absence of comorbid conditions does not reflect medical reality—particularly among geriatric patients. Yet, since the treatment is supported by the evidence, there is pressure for third-party payers to cover its cost. This pattern is true for both private insurance as well as for government-based programs such as Medicare.

1.9 Conclusion

While many of these technology-based end-of-life decisions are not yet common-place in developing countries, as the number of intensive care units in Asia and Africa increases along with medical devices such as ventilators, these concerns are likely to arise. Moreover, the Westernization of medicine worldwide brings new ethical issues and medico-legal standards to developing countries. As noted by Blank (2011), 75% of the world's population resides in regions in which disclosure of medical "bad news" and autonomous patient decision-making are not the norm nor seen as desirable. However, it is often implied that the Western emphasis on patient autonomy is the standard that all nations should adopt in their health care systems. However, as illustrated by the Navajo and Mrs. Kim, even in the U.S., this emphasis on isolated, self-determination may not be a universally-held value.

References

Beckstrand, R. L., Callister, L. C., & Kirchhoff, K. T. (2006). Providing a "good death": critical care nurses' suggestions for improving end-of-life care. *American Journal of Critical Care, 15*(1), 38–45.

Blackhall, L. J., Murphy, S. T., Frank, G., Michel, V., & Azen, S. (1995). Ethnicity and attitudes toward patient autonomy. *JAMA, 274*(10), 820–825.

Blank, R. H. (2011). End-of-life decision making across cultures. *The Journal of Law, Medicine & Ethics, 39*(2), 201–214.

Carrese, J. A., & Rhodes, L. A. (1995). Western bioethics on the Navajo reservation: Benefit or harm? *JAMA,* 826–829.

Combs, S. A., Culp, S., Matlock, D. D., Kutner, J. S., Holley, J. L., & Moss, A. H. (2015). Update on end-of-life care training during nephrology fellowship: A cross-sectional national survey of fellows. *American Journal of Kidney Diseases, 65*(2), 233–239.

Constand, M. K., MacDermid, J. C., Dal Bello-Haas, V., & Law, M. (2014). Scoping review of patient-centered care approaches in healthcare. *BMC Health Services Research, 14*(1), 271.

Diem, S. J., Lantos, J. D., & Tulsky, J. A. (1996). Cardiopulmonary resuscitation on television—miracles and misinformation. *New England Journal of Medicine, 334*(24), 1578–1582.

Emanuel, L. L., Barry, M. J., Stoeckle, J. D., Ettelson, L. M., & Emanuel, E. J. (1991). Advance directives for medical care—A case for greater use. *New England Journal of Medicine, 324*(13), 889–895.

Frank, G., Blackhall, L. J., Michel, V., Murphy, S. T., Azen, S. P., & Park, K. (1998). A discourse of relationships in bioethics: Patient autonomy and end-of-life decision making among elderly Korean Americans. *Medical Anthropology Quarterly, 12*(4), 403–423.

Grisso, T., & Appelbaum, P.S. (1998). *Assessing competence to consent to treatment: A guide for physicians and other health professionals*. New York: Oxford University Press.

Kaufman, S. R. (2015). *Ordinary medicine: Extraordinary treatments, longer lives, and where to draw the line*. Duke University Press.

Kennedy, I. M. (1976). The Karen Quinlan case: Problems and proposals. *Journal of Medical Ethics, 2*(1), 3.

Markel, H. (2017). Epic failure. *The Milbank Quarterly, 95*(3), 451–456.

McFadden, R. D. (1985). Karen Ann Quinlin, 31 Dies, Focus of Right to Life Case, New York times, June 12.

Pence, G. (2003). *Great cases in medical ethics* (4th ed.). New York: McGraw-Hill.

Pence, G. (2016). *Medical ethics: Accounts of groundbreaking cases* (8th ed.). New York: McGraw Hill.

Qazi, H. A., Chen, H., & Zhu, M. (2018). Factors influencing dialysis withdrawal: A scoping review. *BMC Nephrology, 19*(1), 96.

Redberg, R., Katz, M., & Grady, D. (2011). Diagnostic tests: Another frontier for less is more: Or why talking to your patient is a safe and effective method of reassurance. *Archives of Internal Medicine, 171*(7), 619–619.

Schlesinger, M. (2002). A loss of faith: the sources of reduced political legitimacy for the American medical profession. *The Milbank Quarterly, 80*(2), 185–235.

Searight, H. R., & Barbarash, R. A. (1994). Informed consent: clinical and legal issues in family practice. *Family Medicine, 26*(4), 244–249.

Searight, H. R., & Gafford, J. (2005). "It's like playing with your destiny": Bosnian immigrants' views of advance directives and end-of-life decision-making. *Journal of Immigrant Health, 7*(3), 195–203.

Tomes, N. (2016). *Remaking the American patient*. Chapel Hill, NC: University of North Carlina Press.

Truog, R. D. (2008). End-of-life decision-making in the United States. *European Journal of Anaesthesiology, 25*(S42), 43–50.

Walter, P. (1997). The doctrine of informed consent: To inform or not to inform. *St. John's Law Review, 71,* 543–590.

Yin, B., Gandhi, J., Limpisvasti, O., Mohr, K., & ElAttrache, N. S. (2015). Impact of fellowship training on clinical practice of orthopaedic sports medicine. *JBJS, 97*(5), e27.

Chapter 2
Ethical Theories Applied to End-of-Life Medical Care

2.1 Models of Ethical Decision-Making

While medical ethics can be dated back to the Ancient Greeks, cotemporary bioethics' origin is typically dated to the early 1970s (Beauchamp & Childress, 2009, 2013). As medicine became more complex with surgical procedures such as organ transplants, advanced life-saving technology including kidney dialysis, mechanical ventilation, and intensive care units, the moral dimensions of healthcare also became more apparent. The Hippocratic Oath was no longer sufficient to guide the types of decisions that healthcare professionals had to make themselves and with their patients. Rothman (1982) notes that the attention to medical ethics led to "strangers at the bedside"— including philosophers who became involved in medical care as members of newly formed hospital ethics committees. There were other strangers appearing at the bedside. While hospitals and other healthcare institutions had historically been overseen by physicians and occasionally by nurse administrators, a new profession, that of healthcare executive became established. These professionals, even though they were not involved directly patient care, rapidly displaced physicians in the organizational leadership of many healthcare institutions. Because ethical decisions could have legal ramifications, the philosophers were sometimes joined by these professional healthcare administrators as well as attorneys specializing in medical law.

However, for healthcare providers seeking a protocol or definitive guide to decision-making about what to tell patients or how to make decisions about maintaining patients on life support, the guidance provided by professional ethicists was often relativistic rather than decisive. It also became apparent with "right to die" legal cases, that the courtroom was probably not the optimal setting to resolve these issues. Additionally, health care ethics and medical law often did not coincide. For example, many physicians, while recognizing the legality of abortion, view the procedure as morally wrong. Physicians and other health care professionals often use terms, "ethical" and "unethical" but may have difficulty articulating the principles and reasoning behind these judgments. The addition of applied philosophers to medical school faculty and hospital ethics committees provided a set of ethical models to use in analyzing moral dilemmas. A brief overview of these theories follows.

2.1.1 Utilitarianism (Consequentialism)

Utilitarianism, generally associated with Jeremy Bentham (1748–1832) and John Stuart Mill (1806–1873), emphasizes the outcome or consequences of actions rather than the acts, themselves (Pence, 2016). At the micro level of individual patients, a correct decision would be an action that had the greatest long-term benefit for the patient. For example, if the patient had an estimated six months to live, a physician might withhold information about the terminal diagnosis so that the patient would not experience emotional distress during the final months of their life. To some extent, the law of double effect reflects a utilitarian perspective. The law of double effect essentially states that if a patient is given a palliative drug to reduce pain and the use of the drug results in shortening the patient's life, typically by being administered in high doses for pain control, the benefit of pain relief may outweigh the added days of a life in pain. This is a common practice and may not be considered direct euthanasia since the objective is to make the patient comfortable rather than to end their life (Boyle, 2004).

While not specifically labeled as such, a utilitarian ethic is operative in cultures in which family members protect patients from information about serious illness (Searight & Gafford, 2005). It may be argued that these families are performing an informal quality-of-life assessment. Their argument is that since the patient's life may be limited and since they are already struggling with a disease which may be associated with significant physical pain, why burden the patient further with emotionally distressing news of a diagnosis such as pancreatic cancer (Searight & Meredith, 2019). The means-to-a-desirable-end implicit in utilitarianism leads to the moral conclusion that withholding information is perfectly acceptable if it results in a desirable consequence. The courts and some ethicists have recognized the value of nondisclosure on the basis of avoiding emotional harm. If, in the larger context of the patient's life, a physician did not notify the patient of information that would be emotionally distressing and this decision reduced the overall lifetime harm experienced by the patient, non-disclosure would be supported by a utilitarian ethic (Searight & Meredith, 2019).

Historically, the courts have been inconsistent in their views of utilitarian or consequentialist perspectives when the rationale for nondisclosure has been to prevent patients' psychological distress. Withholding information may be acceptable if providing the information would make the patient unwilling to go through surgery to correct a life-threatening condition—a situation termed "therapeutic privilege" (Johnston & Holt, 2006).

The rise of a consumer orientation among patients (Tomes, 2016), and a preference in the U.S. for addressing patient dissatisfaction with medical care through the courts, makes the utilitarian argument for withholding information from patients less viable

(Searight & Meredith, 2019). Ethically, the growing significance of principles such as self-determination also make this position difficult to support.

Hardwig's (1990) attention to the impact of an individual's illness on the patient's family also includes some utilitarian reasoning. While he may object to the utilitarian label, Hardwig's (1990) argument is that a patient who has a terminal ongoing illness, who is not likely to recover but whose maintenance on life support requires significant financial time and energy investments on the part of family members, is an ethically imbalanced state of affairs; many are suffering because of one individual's unfortunate medical state. These contractual principles, while perhaps sound from a utilitarian perspective, create a morally uncomfortable "trade relationship" (Hardwig, 1990) between family members. Finally, from a cultural perspective, utilitarianism's future orientation is not congruent with socio-cultural norms prioritizing the present (Hall, 1989).

2.1.1.1 "Big Picture" Utilitarianism in Health Care

In its most direct form, utilitarianism holds that an action by a healthcare professional resulting in the greatest benefit to the greatest number of people is considered to be the ultimately correct act. From a utilitarian perspective, there are no acts that are considered universally right or wrong. One can only evaluate the morality of an act in terms of its consequences (Searight & Meredith, 2019). At a community, state or national level, utilitarianism addresses the optimal use of limited healthcare resources. If a community hospital has a limited number of intensive care unit beds and a patient who has been comatose for eight weeks is continuing to be treated there at the insistence of family, hospital administrators will often intervene and recommend that the patient be moved to a nursing facility. The argument is that ICU beds are limited in number, care on the unit is costly, and this specialized treatment could better serve patients with a greater chance of recovery. While this underlying reasoning is often not stated explicitly, it is a major consideration in making these decisions-particularly in areas where intensive care unit beds are limited.

2.1.1.2 QALYS, Health Care Policy, and Utilitarianism

Utilitarian thinking is implicitly involved in any type of rationing system. While Britain's National Health Service (NHS) often takes issue with the term "rationing" being applied to the reasoning behind their policies, the principle of devoting limited resources such as ventilator support in the hospital or kidney dialysis is based upon the anticipated health benefit to patients with attention to their estimated lifespan. More recently, the NHS has included the index, quality adjusted life years (QALY), as a factor in making utilitarian decisions (Edwards, Crump, & Dayan, 2015). Quality adjusted life years (QALYs) is a quantitative index of disease burden. It is a numerical figure representing both the quality and quantity of one's life. QALYs can be used to represent the impact of an illness as well as the quantitative benefit of specific treatments. Mathematically, in assessing the potential benefit of an intervention such as stent placement, bariatric surgery for weight loss, open heart surgery, or treatment of HIV with antiretroviral therapy, the quantitative value associated with the treatment is multiplied by the duration of the treatment effect. Essentially, how much improvement does the patient exhibit as a result of the intervention and to what extent does the treatment lengthen their life? (Edwards, Crump, & Dayan, 2015). However, the level of improvement associated with a treatment includes a subjective component through expert opinion on the quality of a patient's life.

Some of the criticisms of QALYs are that the numerical indices are not sensitive to small, yet meaningful, changes in functional status associated with conditions such as cancer. Additionally, the index has been seen as biased against the elderly. An intervention that might improve cognitive functioning in an older patient might not register significant change in QALYs if overall lifespan is not extended (Pettitt et al., 2016). Despite this criticism of constructing a quantitative index to reflect something as subjective as one's quality of life, the index is applied in rationing policy decisions such as in Great Britain's National Health Service (NHS).

It had also been hoped by some healthcare policy professionals that QALYs could provide a seemingly "scientific" basis for private or government-based health insurance plans to refuse to fund treatments with a low probability of success. Pettitt et al. (2016) note that among the ten medications in the United States that have the highest sales and generate the most income for manufacturers and shareholders, these drugs only benefit between 4 and 25% of patients taking them. When the question comes to high priced cancer drugs that may only benefit 20–30% of cancer patients, the question arises about whether this is the best use of limited financial health resources-an issue that is particularly prominent in countries with government-based universal coverage.

As mentioned above, the NHS often relies on QALYs in determining the funding of treatments. In addition, archival data are often used to compare effectiveness of newer treatments to existing therapies in terms of QALYs. However, the economic costs of treatment are significantly weighted in some calculations.

A recent news article, stated that: "The priority ... [of the NHS] has changed from providing all the care that patients need... to dividing up the care that is available (or the budget for it) so that it is distributed equitably" (Laurence, 2015). The organization within the British government that makes these decisions is the National Institute for Health and Care Excellence (NICE). A QALY standard used by NICE is that a proposed treatment must provide improved quality or length of life at a cost of 20,000–30,000 English pounds.

A previously set limit for patients for end-of-life care was 50,000 English pounds per QALY but may be as high as 80,000 English pounds for terminally ill patients "...for whom an extra few months means more than for other patients" (Laurence, 2015).

The government of Singapore is performing a utilitarian analysis of QALYs as applied to end-of-life care, and has been examining several policy alternatives. Singapore's MediShield is a health plan for older adults which historically was for catastrophic health problems. However, as the population of the country ages, there has been interest in extending the range of benefits within this plan. In a study examining the public's preferences for end-of-life treatments, Finkelstein, Bilger, Flynn, & Malhotra (2015), in comparing older community residents with similarly aged patients with cancer, found differences in types of treatment for which potential patients were willing to pay. Those with cancer prioritized pain reduction and being able to die at home. Of note, treatments that might provide moderate life extension were not as important as high quality palliative care. From a fiscally driven utilitarian perspective, the investigators concluded that future healthcare reform in Singapore should not overemphasize costly end-of-life treatment since this was not the priority of its senior citizens (Finkelstein et al., 2015).

In the 1980s, Britain's NHS reportedly set a limit on kidney dialysis and trans-plants—recommending these procedures only for patients 50 or under (Berlyne, 1982). An international study comparing nephrologists' referral patterns for dialysis found that American kidney specialists were more likely than Canadian or British nephrologists to offer their patients dialysis (McKenzie, Moss, Feest, Stocking, & Siegler, 1998). The investigators suggested that the American nephrologists were more motivated by patient and family requests for treatment as well as fear of law-suits than those practicing in Canada and Great Britain—both of which have universal national health coverage. This difference was particularly pronounced when exam-ining patients in a persistent vegetative state and those who exhibited dementia. In these circumstances, almost 50% of the American nephrologists indicated that they would provide dialysis if the family insisted upon it—rates for Canadian kidney spe-cialists were 30% with 16% of British nephrologists agreeing to the treatment with family insistence. About 10% of Canadian and British kidney specialists indicated that financial constraints played a role in withholding dialysis (MacKenzie et al., 1998). The NHS website for patients who are candidates for dialysis indicates that someone who is in their 20s when they begin dialysis can expect to live 20 more years; however, for those over 75 who begin dialysis, the site indicates that their life may be prolonged for only 2–3 years.

2.1.2 Kant's Deontology

The term deontology emphasizes duty—our behavior towards others is rooted in our obligations. An essential Kantian concept is the categorical imperative–there are moral absolutes that exist apart from the social context and laws. While there may be situations in which the ends justify the means, the Kantian worldview holds that the motivation for an act is irrelevant. For example, failing to disclose a cancer diagnosis to a frail 92 year old, since this knowledge will create emotional distress, while not significantly altering the patient's life course, is deemed morally wrong since dishonesty and deception are involved. A physician who does not routinely disclose caner diagnoses to patients ("You have a small mass on your bladder; we can take care of it.") would be judged not on their benevolent motivation for obscuring the truth to protect the patient from emotional harm but solely upon the fact that their actions violated the categorical imperative of honesty. Deception is not permissible under any circumstances regardless of honesty's impact on the patient (Searight & Meredith, 2019). A Kantian principle of always telling the truth would not be well-received by the Navajo- and Korean-American interviewees described in Chapter One. While Kant would recognize that the motivation for withholding information may be to spare the patient harm, the consequences of being less than forthcoming do not permit being untruthful (Searight & Meredith, 2019).

Physicians who do not accept assisting patients in ending their lives under any circumstances—no matter how much pain and suffering the patient experiences, the limited lifespan they have, the impact on the patient's family etc.—are acting from a Kantian categorical imperative. The physician's role is to preserve life rather than expedite a "good death." Hardwig (2000) criticizes this approach as being dehumanizing—there are no people, simply rules. Moreover, Kantian ethics denies motivations and intentions in favor of impartiality (Hardwig, 2000).

Kantians also believe that people have inherent worth and that human beings should never serve as a means to an end. This principle is raised when a patient who has suffered "brain death" is viewed as a source for organs that would benefit other patients. A process of implicit rationing of intensive care unit resources in favor of younger patients with fewer comorbid medical conditions over elderly patients in overall ill health would violate the Kantian precept that no one's life is more valuable than that of another person. Kant would likely find the use of QALYs to decide on treatments for specific patients to be abhorrent. While many physicians see medicine as too complex to have universal principles, moral absolutes do become evident around issues such as euthanasia. Most countries, while permitting withdrawal of life support, view deliberate physician acts to end a patient's life, such as intentionally prescribing lethal doses of medication, as homicide.

2.1.3 Virtue Ethics

Virtue ethics, exemplified by the Hippocratic Oath, (e.g., "Into whatever houses I enter, I will go into them for the benefit of the sick and will abstain from every voluntary act of mischief and corruption; and further from the seduction of females or males, of freemen or slaves") emphasizes the physician's moral character. In contemporary medicine, the Oath has often been reduced to "first do no harm," despite the fact that Hippocrates, did not write this famous phrase as part of the Oath. Hippocrates did indicate that a virtuous physician would "take care that …[patients]…suffered no hurt or damage.".

With respect to assisting in an expedited death at the end-of-life, the Hippocratic Oath states "I will neither give a deadly drug to anybody if asked for it, nor will I make a suggestion to this effect." However, several ethicists have noted that Hippocratic physicians would not prolong a natural process of dying with interventions when death was relatively imminent and unavoidable (Kass, 1991). However, Hippocrates viewed the virtuous physician as one that would protect the patient from "needless" emotional distress. Conceivably, Hippocrates would agree with cultures that believe that disclosing diagnostic and treatment information to patients can be harmful. Yet, Hippocrates, while limited in his choice of medical interventions, did seem to individualize care based on patient characteristics.

In the early 1900s, Robert Cabot (1868–1939,) a member of the Harvard medical school faculty and a largely self-educated ethicist, addressed the shift in views of an effective modern physician. Rather than character traits, Cabot argued that, with the rise of hospital care and scientific discoveries, physicians should "understand the specific diseases, their causes, signs, symptoms, courses, prognoses, treatments—and whether each practitioner applied this understanding in the assessment and management of each individual patient" (Jonsen, 2000, p. 85).

The rise of specialty medicine redefined the qualities of a good physician; "the all-purpose general practitioner of the 19th century-a Christian gentleman of intrinsic goodness, law-abiding and loyal to the codified rules of one professional society" was no longer the ideal (Jonsen, 2000). "Whether or not the practitioner went to church on Sunday, knew the Star-Spangled Banner, swore the Hippocratic Oath, or adhered to precepts about consultation in the AMA code of ethics were …[no longer] important criteria for judging professional propriety." (Burns, 1977; cited in Jonsen, 2000, p. 85).

Virtue ethics shares with Kant's deontology an emphasis on the clinician's acts rather than their consequences for the patient. The virtue perspective often takes the form of a list of the actions and motivations of a "good" person. In the case of healthcare, virtue ethicists, when confronted with a moral dilemma, ask "What would a good physician do?" The Hippocratic Oath reflects this perspective as a narrative of desirable and moral behaviors that an ethical physician exhibits.

In contemporary medical ethics, Pellegrino and colleagues (e.g., Pellegrino & Thomasma, 1987) have been the most visible proponents of the virtue perspective. A virtuous physician is one who embodies characteristics such as intellectual honesty, compassion, fortitude, temperance, integrity, and self-effacement. In dealing with patients, a virtuous physician carries out their duties with benevolence and humility. A virtuous physician is also reflective and acts from a position of prudent wisdom, termed *phronesis* (Pellegrino & Thomasma, 1987; Searight & Meredith, 2019). The virtues perspective views end of life care as fundamentally guided by the physician's fiduciary responsibility to the patient rather than hospital procedures or legalisms. Even though a medical action is the "right thing to do," duties cannot be isolated from the physician's overall character. Virtue ethics, while certainly considering the consequences of physician actions or recommendations, is not utilitarian. However, this perspective is not strictly deontological, either. In contemporary medicine, adherents of virtue ethics attempt to protect the patient from harm while encouraging and representing patients' independent decisions.

True beneficence typically involves fidelity to the best interests of the patient and also results in enhanced patient autonomy (Pellegrino & Thomasma, 1993). However, recent discussions of informed consent suggest that full disclosure of medical information may not necessarily enhance patient autonomy, particularly among those who become anxious and are prone to cognitive distortions when confronted with personally-relevant health care decisions. In reality, potentially threatening information may reduce autonomy through eliciting anxiety, which in turn, compromises rational information-processing (Epstein, Korones, & Quill, 2010). In situations where the physician is reluctant to disclose an ominous prognosis, adherents of virtue ethics encourage the clinician to consider what a trusted physician mentor would do in the same situation (Searight & Meredith, 2019).

2.1.4 Principlism

Beauchamp and Childress (2009) reviewed existing philosophical approaches to medical dilemmas and distilled them into four key principles: autonomy, benefi- cence, non-maleficence, and justice. Principlism, sometimes referred to as "the four principles" is the dominant approach to medical ethics in the United States. The first three dimensions are most relevant in clinical situations; justice (i.e., treating others equally) is relevant when end-of life decisions impact others directly, as in family members quitting their jobs to care for a relative in a persistent vegetative state or at the level of society, or when aggressive late life care contributes to inequitable distribution of scarce health care resources.

Autonomy includes the concept of personal independence; the ability to exert control over important decisions impacting one's own life. The use of advance direc- tives, as encouraged by the PSDA, allows one's own independent values about the quality of life and type of medical treatment desired to be expressed even when the patient can no longer communicate. Patient autonomy is violated or compromised when individuals have not been given all relevant information about a medical deci- sion or are unable to cognitively retain and assimilate information and apply it to their current health circumstances. A major criticism of principlism that makes it difficult to apply to end of life dilemmas—is that the principles may conflict with one another. When those conflicts arise, there is no logical system to resolve these contradictions.

However, many commentators have indicated that the four principles are, in reality, not equally weighted in patient decision-making and that autonomy is "first among equals" (Gillon, 2003). While Beauchamp and Childress (2009), the major propo- nents of principlism, assert otherwise, perusal of legal rulings and cases addressed by hospital ethics committees, finds that patient autonomy is a common nexus of conflict. This assumption of equality among the principles also becomes question- able when viewed in the context of some of the major controversies in bioethics in the past 20 years. Issues such as physician assisted suicide, informed consent, the right to die, advance directives, and patient competence for medical and financial decision-making, all involve patient autonomy as a central value.

Non-maleficence, doing no harm, is often invoked in the controversy surrounding physician-assisted death (PAS). Physicians who view their role as "healers," who must always strive to benefit patients, see assisting with death as violating the Hip- pocratic Oath but also as conflicting with the physicians's vocation to prevent harm (non-malificence). Physcians who participate in expediting a patient's death taint the profession.

Beneficence, the moral obligation to help others, has often been fused with non-maleficence (Beauchamp & Childress, 2009). Nearly all medical interventions have associated risks. In making the decision about a surgical hip replacement for a 92 year old man with heart failure, the benefits cannot be assessed without attention to the potential harm.

The recent right-to-die case of Brittany Maynard illustrates the priority given to autonomy. Ms. Maynard was a 29-year-old woman with a Master's degree in education who had previously taught in several Asian countries. In January 2014, Ms. Maynard was diagnosed with a form of brain cancer. While surgery was performed, the cancer returned in April of that year and she was given a prognosis of six months to live. Shortly afterwards, Ms. Maynard moved to Oregon which has a "death with dignity law" permitting physician assisted expedited death. Ms. Maynard had recently married, was under age 30, and was described as a lively and caring young woman. Her relative youth and frequent articulate media appearances made her different from most patients seeking physician-assisted death. During the last months of her life, Ms. Maynard spoke publicly about the right to die and encouraged other states to pass legislation similar to that in Oregon (Maynard, 2014). Some commentators in the media openly criticized Ms. Maynard for her promotion of a patient's right to end their own life. Within days of her death at age 29, the Catholic Church reiterated their position that physician-assisted death was inconsistent with Church teaching and demonstrated a disrespect for life.

During the months before her death, Ms. Maynard publicly stressed the importance of her control over the circumstances of her death. From the perspective of nonmaleficence, Maynard did not want to compound her death with further physical degradation and also did not want to experience an impaired quality-of-life. Maynard and those who supported physician-assisted death also viewed it as a beneficent act–a merciful way to avoid the prolonged death that has become common in U.S. health care. Angell, a former editor of the New England Journal of Medicine, suggests that recent polls of physicians indicating that the majority are against physician-assisted death may reflect deeper motives. Angell (2014) suggests that discussions of physician assisted death elicit a sense of helplessness when health care professionals confront the limits of medical science in effectively managing patient suffering at the end-of-life.

The weakness of principlism, the absence of a system to resolve differences between competing principles, is also one's of its assets. These four principles can be readily adapted to morally-based medical dilemmas arising internationally and cross culturally (Gillon, 2003). Indeed, principlism is becoming the "preferred" approach to medical ethics across the globe.

2.1.5   Communitarianism

Communitarianism is not a singular ethical theory. A shared element in communitarianism is that morality does not develop in isolation and is only acquired through socio-cultural interaction. Azzi (nd) describes language as the basis for shared communities which generate consensually held moral standards.

Communitarianism usually reflects a dialectic between individual well-being and that of the greater community. Individual worth cannot be viewed apart from the socio-cultural matrix in which one lives. Additionally, individual rights tend to be deemphasized in place of an interdependent web of responsibilities to others (Christians, 2004). Two examples of ethical models that include communitarian themes, Ubuntu ethics from Africa and Confucian ethics from China are discussed below.

2.1.5.1   African Medical Ethics

At the outset of this brief summary, it is important to recognize that, there is no overarching African culture with its own set of ethical guidelines. Additionally, cultural belief systems differ by region—Eastern, Central, Western, and Southern Africa have distinctive groups of cultures. Overall, Africa is estimated to have between 800 and 1200 distinct cultural groups. Additionally, religion influences African health care ethics with both Islam and Christianity becoming increasingly influential.

Some ethicists, however, describe "Pan-African" ethical models such as Ubuntu (Chuwa, 2014). Ubuntu, which has been translated as "humanity toward others" (Christians, 2004) is characterized by an inherent tension between the individual and community. While everyone has basic rights and dignity, individual rights and well-being depend. upon the well-being of the community. Human rights do not exist outside of one's immediate cultural community. As is the case with other non-Western ethical traditions, there is an ongoing tension or dialectic between individual autonomy and the well-being of the community (Chuwa, 2014). While Western bioethics has a set of finite principles such as Kant's' categorical imperative or utilitarianism's perspective of consequentialism, these dichotomous, either-or, moral priorities are inconsistent with ethical systems such as Ubuntu.

Instead, Ubuntu is a "both and" ethical model rather than a Western influenced "either – or" theory (Chuwa, 2014). Individual rights are meaningless if artificially removed from the community of which one is an interdependent part. Individual development and well-being can only occur within a community context. A person is a moral and ethical agent only when they give adequate attention to obligations and responsibilities to others. In keeping with this perspective, healthcare decisions are not person centered since the individual's health influences the community and vice versa. This indivudal-community dualism has led to conflicts when members of African cultures such as the Yoruba receive healthcare in Western settings. Important decisions about matters such as life support would not be made until others are convened (Jegede & Adegoke, 2016).

Living wills and advance directives, while beginning to gain some recognition in Africa, are culturally incongruent with communal agreements and/or ultimate decisions made and communicated by an elder. However, the elder does not fill their role as the spokesperson until after group discussion (Jegede & Adegoke, 2016). To a Western physician, this approach may seem inefficient. However, as is the case with many communitarian perspectives, it is to important consider the impact on important others of a seemingly individual health care decision. While this approach may seem paternalistic, from the perspective of principlism, communitarians typically elevate beneficence relative to autonomy (Kasenene, 2000).

2.1.5.2 Confucianism

Confucian ethics begins with an emphasis on relationships. At the same time, however, the development of virtues is also a component. Confucian ethics are very practical and address many of the moral dilemmas arising in everyday life. In addition to emphasizing benevolence, Confucian ethics also includes the concept of shame. Shame has a role in the origination of morality, since during the course of human development it can develop into *yi* or righteousness. While some ethicists have suggested that Confucian ethics disavows the concept of autonomy, the emphasis on virtues highlights the importance of individual moral character. In Confucianism, the virtuous individual lives according to principles of right and wrong based on an internalized set of values that are maintained in the face of challenges to the virtuous life.

A central concept in Confucian ethics is *Ren*–best described as warm compassionate feelings for others (Yang, 2015). The importance of *Ren* prioritizes the family above the individual. Families cannot be reduced to groups of individuals. Indeed, there is a concept of family as united into one—"one flesh." The emphasis on *Ren* is accompanied by the accompanying concept of filial piety, respect and responsibility for one's family—particularly for parents and grandparents. Filial piety means that when a family member falls ill, the elder's sickness is a collective experience. From the Confucian perspective, the physician actually joins the family. Wang (2015) describes how the family preserves the patient's personal identity and integrity while they are interacting with physicians and the medical system; "… the familial relation is distinguished from other relations in that the family members are committed to sharing each other's fate from birth to death and this initial commitment is unconditional…" (Wang, 2015, p. 77).

Xiao, devotion to one's parents, is the foundation *of. Ren*, which, in turn, is founded upon both filial piety (*Xiao*) and brotherly love (*Ti*). *Xiao*, when in relation to parents, includes *jing* or reverence for one's mother and father. This reverence is imbued with a strong sense of dignity. If adult children only "go through the motions" of providing physical care in the absence of this respectful reverence, the expression of *Xiao is* not genuine (Yang, 2015). Even when parents are behaving wrongly, Confucius urges adult children to maintain their reverence and gently redirect their parents to more appropriate behavior (Yang, 2015).

Confucianism exerts a significant influence on end-of-life care in mainland China, Taiwan and Korea, but is significantly less influential in Japan. Buddhism and Shintoism influence Japanese perspectives (Cheng et al., 2015). Even in Korea, 40% of the population identifies as Buddhist or Christian. Additionally, Buddhism and Taoism are influential in Taiwan. Confucianism, as a philosophy, however is also influential in these countries (Cheng et al., 2015).

2.2 Conclusion

This chapter has been a very brief, limited overview of selected ethical theories in medicine that are applied to end-of-life decision-making. Hopefully, it provides a set of frameworks for analyzing ethical decision-making across nationalities and cultures. When the question arises, "Is that ethical," The response should be "Ethical according to which philosophical theory?" While this chapter also reflects the Western bias in bioethics by emphasizing ethical models developed in Europe and the United States, these approaches, particularly principlism, are being imported into Asia and Africa. From the perspective of respecting diversity and the richness that alternative non-Western schools of thought provide, it is genuinely hoped that these Western models do not supplant theories such as Confucianism and Ubuntu. In a later chapter, focusing on the role of religion, ethical principles deriving from Christian, Judaic and Islamic religions and their views of end-of-life care will be addressed.

References

Angell, M. (2014, October, 31) The Brittany Maynard effect: How she is changing the debate on assisted dying. *Washington Post.*

Beauchamp, T. L., & Childress, J. F. (2009). *Principles of biomedical ethics* (6th ed.). New York: Oxford University Press.

Beauchamp, T. L., & Childress, J. F. (2013). *Principles of biomedical ethics* (7th ed.). New York: Oxford University Press.

Berlyne, G. M. (1982). Over 50 and uremic= death. The failure of the British National Health Service to provide adequate dialysis facilities. *Nephron, 31*(3), 189–190.

Boyle, J. (2004). Medical ethics and double effect: The case of terminal sedation. *Theoretical Medicine and Bioethics, 25*(1), 51–60.

Burns, C. R. (1977). Richard Clarke Cabot (1868–1939) and reformation in American medical ethics. *Bulletin of the History of Medicine, 51*(3), 353–368.

Cheng, S. Y., Suh, S. Y., Morita, T., Oyama, Y., Chiu, T. Y., Koh, S. J., ... & Tsuneto, S. (2015). A cross-cultural study on behaviors when death is approaching in east asian countries: What are the physician-perceived common beliefs and practices? *Medicine, 94*(39), 1–5.

Christians, C. G. (2004). Ubuntu and communitarianism in media ethics. *Ecquid Novi: African Journalism Studies, 25*(2), 235–256.

Chuwa, L. T. (2014). *African indigenous ethics in global bioethics: Interpreting Ubuntu* (Vol. 1). Springer.

Edwards, N., Crump, H., & Dayan, M. (2015). Rationing in the NHS. *London: Nuffield Trust, 8.*

Epstein, R. M., Korones, D. N., & Quill, T. E. (2010). Withholding information from patients—When less is more. *New England Journal of Medicine, 362*(5), 380.

Finkelstein, E. A., Bilger, M., Flynn, T. N., & Malhotra, C. (2015). Preferences for end-of-life care among community-dwelling older adults and patients with advanced cancer: A discrete choice experiment. *Health Policy, 119*(11), 1482–1489.

Gillon, R. (2003). Ethics needs principles—Four can encompass the rest—And respect for autonomy should be "first among equals". *Journal of Medical Ethics, 29*(5), 307–312.

Hall, E. T. (1989). *The dance of life: The other dimension of time.* New York: Anchor.

Hardwig, J. (1990). What about the family? *Hastings Center Report, 20*(2), 5–10.

Hardwig, J. (2000). *Is there a duty to die?: And other essays in bioethics.* Routledge.

Jegede, A. S., & Adegoke, O. O. (2016). Advance directives in end of life decision-making among the Yoruba of South-Western Nigeria. *BEOnline: Journal of the West African Bioethics Training Program, 3*(3), 41.

Johnston, C., & Holt, G. (2006). The legal and ethical implications of therapeutic privilege-is it ever justified to withhold treatment information from a competent patient? *Clinical Ethics, 1*(3), 146–151.

Jonsen, A. R. (2000). *A short history of medical ethics.* New York: Oxford University Press.

Kasenene, P. (2000). African ethical theory and the four principles. *Principles of Healthcare Ethics,* 347–357.

Kass, L. R. (1991). Why doctors must not kill. *Commonweal, 118*(14), S8.

Laurence, J. (2015, March). The cost of NHS health care: Deciding who lives and who dies. *Independent.*

Maynard, B. (2014). My right to death with dignity at 29. *CNN. November,* 2.

McKenzie, J. K., Moss, A. H., Feest, T. G., Stocking, C. B., & Siegler, M. (1998). Dialysis decision making in Canada, the United Kingdom, and the United States. *American Journal of Kidney Diseases, 31*(1), 12–18.

Pellegrino, E. D., & Thomasma, D. C. (1987). The conflict between autonomy and beneficence in medical ethics: Proposal for a resolution. *Journal of Contemporary Health Law and Policy, 3,* 23–46.

Pellegrino, E. D., & Thomasma, D. C. (1993). *The virtues in medical practice.* New York: Oxford University Press.

Pence, G. (2016). *Medical ethics: Accounts of groundbreaking cases* (8th ed.). New York: McGraw Hill.

Pettitt, D. A., Raza, S., Naughton, B., Roscoe, A., Ramakrishnan, A., Ali, A., ... & Brindley, D. A. (2016). The limitations of QALY: a literature review. *Journal of Stem Cell Research and Therapy, 6*(4).

Rothman, D. J. (1982). *Strangers at the bedside: A history of how law and bioethics transformed medical decision making.* New York: Basic.

Searight, H. R., & Gafford, J. (2005). "It's like playing with your destiny": Bosnian immigrants' views of advance directives and end-of-life decision-making. *Journal of Immigrant Health, 7*(3), 195–203.

Searight, H. R., & Meredith, T. (2019). Physician deception and telling the truth about medical "Bad News": History, ethical perspectives, and cultural issues. In *The Palgrave Handbook of Deceptive Communication* (pp. 647–672). New York: Palgrave-Macmillan.

Tomes, N (2016). *Remaking the American patient.* Chapel Hill, NC: University of North Carolina Press.

Wang, J. (2015) Family and autonomy: Towards shared medical decision-making in light of Confucianism. In R. Fan (Ed.), *Family-oriented informed consent: East Asian and American perspectives* (pp. 63–80). New York: Springer.

Yang, Y. (2015). A family oriented Confucian approach to advance directives in end-of-life decision-making for incompetent elderly patients. In R. Fan (Ed.), *Family-oriented informed consent: East Asian and American perspectives* (pp. 257–270). New York: Springer.

Chapter 3
A History of Physician "Truth Telling," Informed Consent, Legal and Religious Perspectives on End-of-Life Care

3.1 Brief History of Physician Disclosure of Diagnostic and Prognostic Information to Patients

While patients' and family preferences for nondisclosure of life-threatening illness found among contemporary Native Americans and Asian Americans may seem to be a cultural anomaly, a brief review of physician practices and legal rulings in the 20th century suggests that disclosure of this information to patients is a relatively recent practice. The shift from paternalistic medicine to collaborative care in which the patient makes the ultimate decision about their treatment became evident in the United States by the 1960s and has accelerated since that time.

While there has always been some disagreement within the medical community about informing patients of a serious illness, there is evidence that in the early 1900s, failure to disclose as well as distortion of health information was common in physician–patient interactions. This practice of withholding medical information was generally motivated by paternalistic beneficence on the physician's part. The view that patients could not psychologically handle medical "bad news" was the chief reason expressed by physicians for this practice.

In the 1920s, several physicians wrote articles for popular magazines in which they openly stated that lying to patients about a serious medical condition was in the patient's best interest. Some physicians such as Collins (1927) went as far as to say that being dishonest with patients was a form of compassion and encouraged his medical colleagues to "...cultivate lying as a fine art" (p. 327).

Cabot (1918, 1926) however, argued that honesty was a preeminent value in medicine. He was disturbed about the extent to which benevolent patient deception had become accepted practice. Collins described instances in which patients directly asked him for diagnostic information who, upon receipt of this information, became very emotionally distressed. Collins (1927) implied that their resulting psychological state hastened their death, Cabot presented an alternative perspective indicating that

© The Author(s), under exclusive license to Springer Nature Switzerland AG 2019
H. R. Searight, *Ethical Challenges in Multi-Cultural Patient Care*,
SpringerBriefs in Ethics, https://doi.org/10.1007/978-3-030-23544-4_3

lying to patients was more likely to be harmful than beneficial. Cabot also argued that communicating honestly with patients actually enhanced their confidence in the physician (Cabot, 1918, 1926; Jonsen, 2007).

While physicians were debating this issue, the general public had not really registered their voice in the discussion. In the early 1950s, Dr. Otis Bowen, who went on to become Secretary of Health, Education and Welfare in the Reagan administration, surveyed nearly 500 patients by mail. He found that nearly all of them (96.5%) wanted to be informed of a cancer diagnosis and 88.5% wanted the physician to inform family members (Bowen, 1955; Searight & Meredith, 2019).

The issue of deception was highlighted in a now classic *New England Journal of Medicine* article by Henry Beecher. An anesthesiologist, Beecher, had reviewed many of the classified military documents about concentration camps for the U.S. Army. Beecher articulately criticized the utilitarian perspective that viewed harm to an individual's health for the greater good of scientific progress as unacceptable. Similarly, experimenting on one patient for the possible benefit of many was a highly questionable practice (Kopp, 1999); the ends did not justify the means. Beecher argued that research of this type could only be conducted with complete consent of the subject/patient involved. Beecher became well-known when in 1965, he submitted a manuscript to the *New England Journal of Medicine* outlining over 20 examples of published medical research that did not meet ethical standards (Beecher, 1966). Not obtaining informed consent was a common problem that he cited in his review of these studies. Some of the experiments reviewed involved children. In at least one study, children with developmental disabilities were unknowingly exposed to radioactive material in their morning oatmeal (Beecher, 1966; Rothman, 1982).

From the perspective of ethical theory, Beecher was clearly in the corner of Kant and his categorical imperative. Informed consent was a fundamental principle that should govern all medical research. He rejected the utilitarian argument that possible harm to an individual as a research participant might lead to medical knowledge that would benefit many patients. This utilitarian argument was one that was used by physician–researchers who performed often gruesome experiments on Nazi concentration camp inmates (Schultz, 2013).

It was not until the 1970s that the field of bioethics, dominated by professionals with a background in philosophy, began to influence medical practice. Elizabeth Kubler-Ross's research on stages of patients' response upon learning of a terminal diagnosis, put death and dying into medical school curricula. Her work attempted to "normalize" death as a universal developmental transition. However, even Kubler-Ross expressed reservations about directly informing patients of a terminal condition. When directly asked by a patient if they were going to die, Kubler-Ross encouraged non-specific responses such as "It does look bad." (Kubler-Ross, 1969; Kubler-Ross, Wessler, & Avioli, 1972).

Outwardly, the change in physician norms around disclosure appeared to be well-established by the end of the 1970s. Two studies on physician disclosure of a cancer diagnosis separated by less than 20 years, illustrate the change in "truth telling" in medicine. In 1961, results of a survey of U.S. oncologists indicated that over 90% had deliberately withheld information about a cancer diagnosis at some time during their medical career (Oken, 1961). In a survey conducted in the late 1970s, 97% of physicians indicated that disclosure of a cancer diagnosis was their usual practice (Novack et al., 1979).

However, anecdotal accounts from the 1970s suggested that at least in some countries, protecting the patient from distressing medical information was still the norm. One of the founders of the "antipsychiatry movement," British psychiatrist R. D. Laing (1927–1989), best known for his book, *The Politics of Experience*, elicited turmoil by telling his young adult daughter of her impending death. In 1976, Susan was hospitalized and dying of recent onset leukemia. Laing, against the wishes of Susan's mother, the nursing staff, Susan's siblings, and the hematologist in charge of her care, disclosed to her that she would soon die. In keeping with his critical stance towards the medical establishment, Laing addressed this "conspiracy of silence" with characteristic bluntness:

> The nursing sister said, "Dr. Laing, you're not going to tell her." "Fuck off." I told her. I just told [Susan] the facts, as I knew them, and she elected to be disconnected and taken back to her boyfriend's flat.…She died in (her boyfriend's) flat and she was glad she had been told. I mean, she said that she was very glad that I told her and thanked me very much for telling her. It divided the family, I mean murder was in the air…the boyfriend said "you've killed my girl, you have destroyed her by telling her.".…He wanted her kept alive for as long as possible in the hope against hope that the clinic might find a cure for it if she was kept alive long enough…[The hematologist] said he'd only told two people in the last 30 years that they are going to die, it's against the policy of the National Health Service in Stobhill to tell people that they're going to die (Mullan, 2017, p. 89).

In the United States, Canada and much of Western Europe, signing the informed consent document became a ritual before any medical procedure—including routine blood draws. In the U.S., during the latter half of the 20th century, medical malpractice litigation rose significantly threatening the financial well-being of hospitals and physician practices. "Risk management" departments became a standard part of health care delivery systems. While there was much greater attention to obtaining signed documents attesting to a patient's informed consent for a medical procedure, this consent was a contractual exercise—often devoid of any genuine education about the procedure, including the procedure's risks and benefits as well as alternatives (Searight & Barbarash, 1994).

Delivering medical "bad news" is stressful for many health care providers. It is likely to elicit strong emotion in the patient and their family which may be threatening for the physician who may feel a sense of helplessness—particularly in cases of terminal illness. This discomfort may be expressed as a hurried informational encounter or by unhelpful, unempathic emotional detachment (Monden, Gentry, & Cox, 2016). Recognizing that physicians experienced discomfort in these discussions, specific training in delivery of "bad news" has become part of medical school and residency education. In keeping with the pedagogic style of medical education, stepwise protocols and acronyms have been developed to assist physicians-in-training in developing these communication skills. For example, the ABCDE model for communicating bad news to patients follows these steps: A—advance preparation, B—build a relationship and establish an appropriate setting; C—communicate effectively; D—deal with the patient's and family's emotional reaction; and E—encourage emotional expression and acknowledge the patient's and family members' feelings (Rabow & McPhee, 1999).

Studies conducted outside the U.S. in the 1980s and 1990s did find that in countries such as Spain, France, and Japan, patients with cancer were rarely told their diagnosis directly and instead, the family was informed. There is evidence that this pattern is changing but these direct disclosures of bad news are not always seen as desirable by health care professionals or the patient's family (Syed, Almas, Naeem, Malik, & Muhammad, 2017). Because of its longer history, U.S. medical ethics is often seen as the standard for other countries. A medical education from a U.S. medical school is valued around the world and it is generally recognized internationally that the United States has the highest level of medical technology available. It is very common for foreign dignitaries to come to United States to undergo cancer treatment or significant surgery. Along with exporting the technical aspects of medical care, concepts such as informed consent, advance directives, autonomous patient decision-making, and more recently physician–patient collaboration, are also being widely taught in medical schools outside of the United States.

3.2 Legal Issues in End-of-Life Decision-Making: Selected International Examples

The legalities around end-of-life care have been addressed in two primary ways. First, some countries and states within the United States have passed legislation that governs issues such as physician-assisted death, advance directives, and proxy decision-making. Second, as has been discussed, end-of-life issues—particularly when family conflict is present—have been addressed through the court system and the rulings often stand as precedents. While many commentators and judges themselves have offered the opinion that the courts are not the optimal place to address issues of this type, the inability of patients, healthcare professionals and patients' families to resolve these issues amongst themselves bring these dilemmas to legal attention. In examining end-of-life legislation and current laws across countries, the most striking observation is their inconsistency. Additionally, laws that have been passed as well as legal cases often have limited applicability to the types of issues arising through societal changes as well as growing medical technology A review of laws governing end-of-life care in selected countries follows.

3.2.1 United States

As noted earlier, the United States enacted the federal Patient Self-Determination Act (PSDA) in the early 1990s. The law has certainly raised the public's awareness of advance directives. Additionally, the absence of a record of the patient's wishes for continued treatment if they were to be in a comatose state was a key element in multiple well-publicized end-of-life cases such as that of Terri Schiavo.

While there are both federal and state regulations regarding advance directives and related issues such as physician-assisted death, much of the legal basis for actions surrounding end-of-life care result from specific court decisions. Many scholars of informed consent law see *Schloendorff vs. Society of New York Hospital* (1914) as the legal basis for patient autonomy. In that now famous ruling, Judge Cardozo concluded that "every human being of adult years and sound mind has the right to determine what shall be done with his own body." If a surgeon performs a procedure without the patient's explicit consent then they have engaged in battery and can be prosecuted. As noted earlier, subsequent court decisions have centered around the amount and clarity of information with which the patient was provided prior to a procedure.

The precedent-setting case for complete informed consent in the U.S. was *Natanson versus Kline*. Ms. Natanson agreed to undergo cobalt radiation therapy for cancer. The treatment resulted in damage to the patient's chest including destruction of some cartilage and bone. In the case, Ms. Natanson argued that the radiologist had not adequately informed her of the risks of the treatment. While U.S. courts have occasionally recognized that it may not be in the patient's best interest to have all available information if that would cause them to refuse treatment for a life-threatening condition, those circumstances did not hold in Ms. Natanson's situation (Searight & Meredith, 2019). The standard of informing the patient of all foreseeable risks of treatment as well as the probable outcomes of not undergoing treatment should be part of obtaining informed consent.

Another example of failure to adequately inform the patient is illustrated by the case, *Kelton vs. District of Columbia*. Ms. Kelton had undergone exploratory surgery to determine why she had been unable to conceive after a previous cesarean delivery. During the exploratory procedure, the surgeon found scarring on the fallopian tubes that appeared consistent with a previous tubal ligation. Ms. Kelton initiated litigation against the physician who had performed the cesarean section but was not successful because of the one-year statute of limitations for battery.

Most healthcare policies involving specific treatments, recognition of advance directives, processes for appointing surrogate decision-makers, and more recently, physician-assisted death are all made at the state level. As a general rule, the federal government and the U.S. Supreme Court have been reluctant to interfere with the states' right to establish health care policy. For example, as of this writing, nine states have legalized physician-assisted death. The first state to do so in the U.S. was Oregon which passed a "right to die" or alternatively, a "death with dignity" law.

3.2.2   Israel

Israeli health law is characterized by a combination of secular and Jewish religious law which may, at times, contradict each other. Additionally, while these two orientations are, in theory, separate, Jewish law exerts significant influence on interpretation and implementation of Israeli secular law.

In 1996, the Israeli parliament, the Knesset, passed a patient's rights law. While the law emphasized patient autonomy and included language somewhat similar to that found in the United States, the Israeli legislation also sets specific limits on patient autonomy. For example, a patient can be treated coercively. This forced treatment has occurred with politically-motivated hunger strikers. The legal rationale for coerced treatment is that life is a definite priority which supersedes individual autonomy and dignity (Sturman, 2003). However, support for compelled treatment against the patient's wishes is not consistent in the Israeli medical community.

A case involving a parental request to discontinue a terminally ill child's life support illustrates how secular and religious law interact. The case involved a child with Tay-Sachs disease. This genetic condition has no cure and death typically occurs by the age of five. As the disease progresses, the child experiences more neurological symptoms including seizures. In this case, the patient's mother had requested that her daughter be permitted to die and requested that medical treatment be discontinued. Since the physician in charge of the case disagreed, the conflict went to court.

The judge in the case argued that both religious law and secular principles needed to be considered and integrated. While noting the importance of autonomy under secular democratic law, any type of euthanasia is not permitted. Additionally, since an absolute value cannot be placed on human life and because of its sanctification, there is no basis for prematurely terminating life. However, the judge recognized that in this particular case, there would be a possible "double effect" in alleviating the child's pain. Thus, the levels of medication needed to satisfactorily address the child's pain could also serve to expedite the child's death. Additionally, while recognizing that some types of life support might be considered extraordinary, the highest value was that of the child's life and even though they were in a terminal state, the child continued to be a sentient being. In her analysis, Sturman (2003) points out that the importance of Jewish religious law in Israel may create inherent conflicts with laws passed by elected representatives. Jewish law would assert that a physician's principal role is to continue to treat terminally ill patients until they are deceased. However, secular law would place greater emphasis on the patient's right to self-determination.

In Israel, the Supreme Court does not have the degree of authority as its U.S. counterpart. One factor contributing to this state of affairs is that Israel does not have a specific constitution (Sturman, 2003). Moreover, the Supreme Court does not have absolute authority in interpreting and applying Israeli law. Court decisions may be modified or overturned by the legislature.

In her detailed account of terminally ill patients in Israeli hospitals, Sturman (2003) is struck by the fact that there are few established written policies and that there is often considerable ambiguity about the appropriate course of action. Much of this diversity stems from different interpretations of Jewish law by rabbinical authorities as well as the physicians understanding of and allegiance to Jewish law. Those healthcare providers who were most heavily influenced by Jewish law were most likely to continue to treat even when, by medical standards, ongoing treatment would be considered futile. Sturman (2003) describes another instance in which the patient had on file a written order for no resuscitation. Even though the physician on call during the night was well aware of the order, they attempted resuscitation anyway.

However, as will be discussed later, physicians in Israel still disagree about the definition of death. While some physicians define death as cessation of heart activity, others use the criterion of the absence of central nervous system functioning.

3.2.3 Europe

Germany: Germany has had national regulation of advance directives since 2009 (Taupitz, 2013). German law includes living wills, powers of attorney and custodianships. Custodianships are implemented for adults who unable to make decisions and manage personal affairs such as finances. In some instances, the person appointed as a custodian may be a relative or family member who is also durable power of attorney. However, the courts may appoint a custodian; court appointed custodians are actually the most common form of custodianship in Germany. Powers of attorney are similar to those in the United States in that they typically involve someone with whom the patient had a long-standing relationship and whom they trust. The person named power of attorney is authorized to make decisions on behalf of the patient. Living wills are somewhat similar to those in the United States in which the patient draws up a written legal document that indicates acceptance or non-acceptance of specific medical interventions if the patient is unable to give consent themselves. Of interest, in the German version of living wills, it is the role of a custodian or power of attorney, rather than the physician, to make the judgment about whether the patient's wishes for medical care are being met (Taupitz, 2013).

The German government has also examined the issue of euthanasia. Any act that is deliberately intended to shorten the patient's life, regardless of the patient's wishes is still considered a criminal act. While within Germany, there has been discussion about active euthanasia as a possibility, the lingering historical impact of Nazi atrocities in which euthanasia was deliberately practiced on segments of the population such as those of Jewish background, appears to be a factor in limiting further legal pursuit of this option.

German legal authorities have also examined the issue of not initiating or withdrawing life-support. In German law, this life support includes artificial feeding or hydration. However, there has been some ambiguity about this issue. In a 2010 case, a 77-year-old woman who was in a nursing home had reportedly been in a persistent vegetative state for five years and was being fed through a stomach tube. It was reported that earlier in her life, the patient had indicated to her children that she did not want to be kept alive in such a state. The physician concluded that continuing artificial feeding was not medically appropriate (Zwick, 2013). The daughter, who was the mother's surrogate, agreed with the decision. However, the supervisory staff at the nursing home overrode the daughter's and physicians plan and continued the feeding. The daughter contacted an attorney who reportedly advised her to remove the feeding tube which she did. However, nursing home staff reinserted the tube. The patient died two weeks later. While the daughter was not found guilty, the attorney was convicted of manslaughter.

Spain: In Spain, regulations regarding advance directives are somewhat confusing which may partially explain the reported low rate of advance directive completion (Simon-Lorda, Tamayo-Velazquez, & Barrio-Cantalejo, 2008). The organization of Spanish government may account for some of this confusion. While there is a national government, Spain is also made up of 17 autonomous regional communities and two cities. Similar to U.S. states, these entities often have the authority to develop regulations around healthcare issues. In some instances, federal and community regulations may not be consistent.

There have been several Spanish cases involving members of the Jehovah's Witness Church whose written advance directives clearly prohibited blood transfusion as in keeping with the tenets of this faith. In at least one of these instances, the judge overruled the patient's advance directive and ordered medical staff to proceed with the transfusion.

While specifying that patient informed consent is important for a valid advance directive, Spanish law also supports a person's right to not be informed of their condition and its prognosis. As is the case with other countries in Mediterranean Europe, physicians are somewhat less likely to disclose medical "bad news" (Gysels et al., 2012). In Spain, there is a clear distinction between an advance directive as a written document and a proxy decision-maker. Patients are encouraged to have a written values history which would include issues such as their perspectives on pain, important family relationships, importance of personal autonomy, religious beliefs and healthcare preferences. This document may be used to guide the decision-making process. It is also recommended that this values history be conveyed to physicians and relatives as well as any official healthcare proxy.

England: Advance care documents are supported in England and Wales. In essence, these documents list treatments that would *not* be acceptable if the individual became incompetent and unable to express their wishes directly. The emphasis has been on reducing futile treatment. Treatment refusals are addressed in *The Mental Capacity Act of 2005*. The Act also specifies that physicians will not be held liable for restricting treatment as specified by the patient's advance refusal document (Lewis, 2012).

Case law in England has focused on withholding and withdrawing treatment in situations in which in the physician's opinion, the treatment was not in the patient's best interests. The affirmative right of physicians to refuse seemingly futile treatment in the United Kingdom is interesting in the context of the National Health Service (NHS). NHS policies have been criticized by some for being guided by utilitarianism—specifically directing medical resources to those who can would be more likely to benefit based on factors such as age and overall health status.

The principle of double effect has also been supported by British courts such that physicians will not be held liable if they provide sedation or pain medication to make the patient comfortable even though it could also likely hasten the patient's death. If the purpose was to deliberately cause the patient's death, the physician would then be guilty of a crime.

Britain has been particularly reluctant to permit euthanasia and physician-assisted suicide (NHS, 2019). Indeed, in one highly publicized case, a husband was prosecuted for accompanying his wife to Switzerland for physician assisted death which is legal in that country. For example, in 2016, British authorities suggested that the wife of a 65-year-old man with multiple sclerosis could be prosecuted for accompanying him to Switzerland where he could receive legal physician assistance in dying through the organization, Dignitas (Finnigan, 2016).

3.2.4 India

Like many developing countries, India does not yet have the medical resources to regularly maintain patients on extended life support. One relevant statute is "Medical Treatment of Terminally Ill Patients" (Protection of Patients and Medical Practitioners Bill). Suicide was decriminalized by statute in India in 2017 after three publicized legal cases addressed the issue without a clear conclusion (Rao, Roy, & Tatiya, 2018).

In a "right to die" case, a nurse in a Mumbai hospital had, in the early 1970s, been assaulted and strangled resulting in serious brain damage. In a persistent vegetative state, she had been residing in the hospital for 36 years. A journalist filed a petition with the courts to end the patient's "suffering "through cessation of feeding. The staff of the hospital who had taken care of the patient for the ensuing 36 years, opposed withdrawal of life support. However, the court, while not permitting withdrawal of support for the patient, did establish guidelines for passive euthanasia and indicated that at present, there was no support for active euthanasia unless appropriate legislation was passed (Rao, Roy, & Tatiya, 2018).

3.3 Religion and the End of Life

Religion is a major moral force throughout the world. Particularly when addressing death, religious belief systems create meaning at the end of a patient's life. While this chapter's description of religion and ethics is not exhaustive by any means, it will, hopefully, provide information about how some of the world's religions view end-of-life care. There actually may be more similarities between religions and how they address end-of-life issues than shared ground between secular models of ethics rooted in philosophy.

Medical ethics in the West is generally discussed from a secular perspective. However, it is important to recognize that the world's religions are founded upon specific moral principles. In most religions, autonomy is recognized but one's life is simultaneously directed by a higher power. This modified autotomy has been termed "theonomy" (Greenberger, 2015). Importantly, rather than addressing moral issues arising in daily life through introspection or discussions with family members or friends, guidance is found in the Scriptures and holy writings of the world's religions (Greenberger, 2015). What one "feels" or thinks is moral is not adequate justification—particularly for serious decisions such as maintaining life support. Instead, the tenets of the individual's religion should be the guide.

Jonsen (2003) states that in the United States, religion exerted considerable influence on American physicians through the early 1900s. However, with the rise in scientism, secularism diminished religion's role in medicine. He states that "by mid-century American medicine had become officially agnostic" (Jonsen, 2003, p. 92). However, this agnosticism is certainly not true around the world and in countries such as Israel, Pakistan and much of the Middle East, end-of-life decisions cannot be made without reference to religion.

Even though the United States has become increasingly secular, Christianity has influenced legislation on topics such as abortion and physician-assisted suicide. In the Blackhall et al. (1995) study, religion influenced European Americans' attitudes towards end of life care with Protestants more likely than non-Protestants to believe that patients should be directly informed of a terminal diagnosis (81 vs. 61%) and with Protestants more likely to believe that the patient should be the primary decision-maker (73 vs. 59%). In the Korean American group, Buddhists were less likely to believe that the patient should be told their prognosis (Blackhall et al., 1995).

3.3.1 Islam

An overarching duty of Muslims is to submit completely to Allah or God. Islam, through the Sharia or religious laws, have a highly systematized set of rules that govern nearly all aspects of life (Al-Bar & Chamsi-Pasha, 2015). The primary basis of Islamic legal opinion is *The Quran*. Another valuable source is *The Sunnah*, providing a description of the life of the Prophet Mohammed and his teachings. If an issue cannot seem to be addressed by either of these sources, Islamic scholars or judges are sought for their opinion. However, these judgments are based on reasoning from the *Quran* or *The Sunnah* (Al-Bar & Chamsi-Pasha, 2015). For example, since it is not explicitly addressed by either the *Quran or The Sunnah*, the opinion of Islamic scholars has been necessary in addressing organ transplantation.

In many Islamic countries such as Pakistan, secular law is informed by Islam. For example, the Constitution of Pakistan states that the country cannot enact a statute considered "repugnant" to the principles of Islam (Moazam, 2000, p. 29). The Islamic Code of Medical Ethics emphasizes the duties and responsibilities as well as the sacred role of the physician while minimizing attention to patient autonomy.

Patients do not have the right to request medical assistance in prematurely ending their life. However, The Islamic Code of Medical Ethics does describe informed consent but also states clearly that in a life-threatening situation, the physician can act without formal consent. While referencing the patient's right to know about their illness, The Code encourages the physician to use terminology that would be minimally distressing to the patient (Moazam, 2000). The physician is a revered figure in the Islamic tradition and healing is a religious act. Some scholars describe the role of physician in Islam as second only to that of Imams and Islamic scholars. The practice of medicine is an act of worship with the physician being an instrument of God (Moazam, 2000). Somewhat similar to the Jewish tradition, from the patient's perspective, illness is something to be accepted and may be a form of atonement. Given that Allah has given the gift of life, only Allah can take it away. As a result, any type of euthanasia or active assistance in dying is forbidden.

The Prophet clearly states that God looks favorably upon those who love their family. The idea of having to develop a contract such as an advance directive is not something that would be even considered (Moazam, 2000) since family members would believe that a loved one's death should not occur prematurely. As noted above, the physician, as a respected healer in Islamic countries, is forbidden from termination of life. Even if the patient requests assistance in dying, the physician should never sanction an act that conflicts with Allah's directives (Nikookar & Sooteh, 2014).

However, while prevention of unnecessary premature death is a physician's duty, when it is clear that the patient will die regardless of any intervention and that no treatment will reverse that imminent end point, treatments may be withheld. However, hastening death is generally seen as forbidden. Similarly, nutritional support should continue to be provided even in patients who are terminally ill and nutrition should never be discontinued as a means of expediting a predicted death. Withholding nutrition would be seen as deliberate starvation which is a criminal act according to Islamic law. However, when death is a clearly predictable and irreversible endpoint, withdrawal of life support may be considered appropriate (Al-Bar & Chamsi-Pasha, 2015).

3.3.2 Judaism

Judaism, similar to most of the world's religions, has ethical principles that unite believers but also encompass multiple interpretations of Jewish law and accompanying practices. However, these principles are rooted in the *Torah* and *Talmud*. Preservation of human life is a fundamental principle in the Jewish faith. The purpose of one's existence or the quality of life do not over-ride or modify the basic respect for all human life (Bleich, 1981, 2006). Questions of whether a life is worth living can only be answered by God. Bleich (1981, 2006) argues that Judaism's emphasis on human life as inherently valuable surpasses that of Christianity. Kant's moral absolutes certainly resonate with this perspective. In Israel, both secular law and Jewish law impact health care decisions. However, within rabbinic law, there are theological disagreements about maintaining life-support and what constitutes death. Sturman (2003) notes that many hospitals in Israel do not have formal policies on end-of-life issues. For example, written do not resuscitate orders are often not available as forms for the medical team to follow. One suggested reason for the absence of consistent policy is that the level of disagreement among health care professionals and rabbinical authorities can be pronounced (Sturman, 2003).

Suffering is part of life and it is never considered to be "needless" or an experience to be avoided. A life of pain and suffering is still considered to be superior to death. Indeed, the gift of life, which only God can provide, outweighs the impact of any tribulations that the patient may endure—even as death approaches. Moreover, given life's sacred quality, physicians have an obligation to be a good steward of this gift and to preserve life. Given this perspective, it is understandable that in Jewish law, abortion is considered immoral. Similarly, any medical intervention that would artificially shorten life, even by a day, would be an insult to God. A close corollary is that one has a strong obligation to act to save the life of a fellow human being. Talmudic literature includes narratives of the necessity of coming to the aid of another such as a person who is near drowning (Bleich, 1981, 2006). The life of a patient in a persistent vegetative state, may have a purpose and should not be extinguished intentionally or prematurely. This purpose may be to teach others sensitivity and compassion (Greenberger, 2015). From the perspective of principlism, Jewish ethics clearly prioritizes beneficence and nonmaleficence over autonomy.

However, if death is believed to occur in the foreseeable future, any obstacle interfering with an overall certain dying process may be removed. A key issue that has arisen with respect to organ transplantation is the criterion for death. Brain death may be considered a definition from the point of view of secular law. However, death from the perspective of Jewish law is the cessation of heart activity. While an absence of brainstem activity is a legal definition for death in Israel, the law allows for families to request continued treatment until cessation of heart activity. As Jotkowitz, Agbaria, and Glick (2017) note, this can lead to the unusual situation of having a patient who is still receiving treatment in an intensive care unit while also having a signed death certificate (Sturman, 2003).

3.3.3 Christianity

Christianity is founded upon the belief that God's son, Jesus, came to earth to redeem humankind. Adherence to the law (largely the Jewish law of the Old Testament) had not been adequate to prevent humankind from straying into sinfulness. Jesus, through his teachings during the course of his lifetime, challenged how Jewish law was being interpreted and how it influenced daily life. Jesus' death is seen as the ultimate atonement and sacrifice for the sinfulness of humankind. His resurrection from the dead demonstrates his divinity while also holding out the promise of life after death for Christians.

Having faith in God and believing that Jesus was both God and Man are central tenets of Christianity. The Bible's New Testament, includes many accounts of Jesus, as God incarnate, being able to work multiple miracles. These included curing the sick, sometimes by laying hands on them and at other times by simply by giving a command from a distance. There are three instances in the New Testament in which Jesus, through his commands, reversed death. In the account of Jairus' 12-year-old daughter Jesus arrives at the home finding that the girl has recently died. However, he tells those already in mourning that the girl is not dead; she is only sleeping. Jesus walks into the girl's room, takes her by the hand and says "Get up" and she does. In another instance, a funeral is already taking place for the son of a widow, Jesus touches the funeral bier and says "Young man get up" and the young man returns to life. Finally, in the case of Lazarus, Jesus' friend and the brother of Mary and Martha, Jesus resurrects him from the tomb. Jesus stands outside of the tomb and in a loud voice says "Lazarus come out!" Lazarus comes out of the tomb bound in burial wrappings.

These miracles in which persons were dead or near death and return back to a normal life are likely to influence Christians' views on issues such as life support and advance directives. In particular, in many African-American churches, the message of God's miracles reflects the superiority of God's work over medical science.

Even in situations involving comatose patients where the likelihood of "awakening" becomes less and less likely as time passes, the fact that this outcome has occurred on rare occasions may be seen as consistent with "the working of God's hands." This message has been used to argue against removing life-support in cases such as that of Nancy Cruzan and Terri Schiavo.

Writing from a secular perspective, Pence (2016) notes that the fact that individuals do come out of multiyear comas "… Changes the prognosis from certainty to probability…." The emotional weight changes when a patient has a 'tiny' chance (Pence, 2016, p. 39). Several examples include Terry Wallace who emerged from a 19-year coma, following an auto accident and Patricia Ingle who went into a "locked in" like state and was unable to speak swallow or move but came back to full consciousness 16 years later (Pence, 2016).

The Catholic Church has addressed the issue of life support. Pope Pius XII in the late 1950s indicated that if there was no reasonable likelihood of recovery, individuals should not be kept alive by extraordinary mean. Ethicists, writing from a Roman Catholic perspective have espoused different perspectives on providing artificial nutrition for those in persistent vegetative states. The view that all life is sacred, even unconscious life, leads very directly to the conclusion that artificial nutrition, which is not considered a treatment but a fundamental human right, should be continued. However, an alternative perspective suggests that the absence of consciousness removes spirituality from one's existence. The absence of a spiritual dimension to human life raises the question about whether this is a genuine human existence. In these circumstances, some discretion around the use of artificial nutrition should be considered (Greenberger, 2015; Ravitsky & Prawer, 2008).

When death is imminently unavoidable, it is acceptable for the physician to refuse to provide treatment that will only lengthen and add discomfort to the process of one's demise. Other Christian denominations are in agreement with the Roman Catholic Church that terminating treatments which will be futile is acceptable. It is however recognized that pain reduction is acceptable even when in some patients it may result in a more rapid death—the law of double effect (Bülow, Sprung, Reinhart, et al., 2008).

3.4 Conclusion

Legal cases addressing end-of-life issues have principally centered around the patient's or patient's families right to continue or cease life support. These more recent cases are extrapolations from earlier rulings that center around the principle that patients should be completely informed before they consent to a medical procedure such as surgery. More recently, the courts have had to address situations in which there is disagreement within the scientific community such as the status of persistent vegetative states and the probability of patient recovery when they are comatose.

Most Western countries have developed legal responses to these questions. The concept of advance directives and proxy decision-makers reflects the contractual nature of the legal system. At the same time, it becomes growingly evident that the legal system has limits in addressing existential questions such as the quality of one's life and even what constitutes life. Religious doctrines do address these issues. As a general rule, Judaism, Islam and Christianity share a respect for life. While there is certainly variability within a religion, there is a shared view that life has inherent value and its outcome is not in the hands of human science.

References

Al-Bar, M. A., & Chamsi-Pasha, H. (2015). *Contemporary bioethics: Islamic perspective*. New York: Springer.

Beecher, H. K. (1966). Ethics and clinical research. *New England Journal of Medicine, 274*, 1354–1360.

Blackhall, L. J., Murphy, S. T., Frank, G., Michel, V., & Azen, S. (1995). Ethnicity and attitudes toward patient autonomy. *JAMA, 274*(10), 820–825.

Bleich, J. D. (1981). Time of death. In J. D. Bleich (Ed.), *Judaism and healing* (pp. 146–157). New York: Klav Publishing.

Bleich, J. D. (2006). "Treatment of the terminally ill,". In P. Hurwitz, J. Picard, & A. Steinberg (Eds.), *Jewish Ethics and the Care of End-of-Life Patients*. New Jersey, NJ, USA: KTAV.

Bowen, O. R. (1955). Why cancer victims should be told the truth. *Medical Times, 83*, 793–799.

Bülow, H. H., Sprung, C. L., Reinhart, K., Prayag, S., Du, B., Armaganidis, A., ... & Levy, M. M. (2008). The world's major religions' points of views on end-of-life decisions in the intensive care unit. *Intensive Care Medicine, 34*(3), 423–430.

Cabot, R. C. (1926). *Adventures on the borderlands of ethics*. New York: Harper and Brothers.

Cabot, R. C. (1918). *Training and rewards of the physician*. New York: JB Lippincott.

Collins, J. (1927). Should doctors tell the truth. *Harper's, 155,* 320–326.

Finnigan, L. (December 10, 2016). Wife could face police investigation after taking her husband to Dignitas flat in Switzerland. The Telegraph. https://www.telegraph.co.uk/news/2016/12/10/wife-could-face-police-investigation-taking-husband-dignitas/.

Greenberger, C. (2015). Enteral nutrition in end of life care: The Jewish Halachic ethics. *Nursing Ethics, 22*(4), 440–451.

Gysels, M., Evans, N., Meñaca, A., Andrew, E., Toscani, F., Finetti, S., ... Pool, R. (2012). Culture and end of life care: A scoping exercise in seven European countries. *PLoS ONE, 7*(4), e34188.

Jonsen, A. R. (2003). *A short history of medical ethics*. New York: Oxford University Press.

Jonsen, A. R. (2007). The god squad and the origins of transplantation ethics and policy. *The Journal of Law, Medicine & Ethics, 35*(2), 238–240.

Jotkowitz, A. B., Agbaria, R., & Glick, S. M. (2017). Medical ethics in Israel—Bridging religious and secular values. *The Lancet, 389*(10088), 2584–2586.

Kopp, V. J. (1999). Henry Knowles Beecher and the development of informed consent in anesthesia research. *Anesthesiology: The Journal of the American Society of Anesthesiologists, 90*(6), 1756–1765.

Kübler-Ross, E., Wessler, S., & Avioli, L. V. (1972). On death and dying. *Journal of the American Medical Association, 221*(2), 174–179.

Kübler-Ross, E. (1969). *On death and dying*. New York: Scribners.

Lewis, P. (2012). The limits of autonomy: Law at the end of life in England and Wales. In S. Negri (Ed.), *Self-determination, dignity, and end of life care*. Leiden: Martius Nijhoff.

Moazam, F. (2000). Families, patients, and physicians in medical decision making: A Pakistani perspective. *Hastings Center Report, 30*(6), 28–37.

Monden, K. R., Gentry, L., & Cox, T. R. (2016). Delivering bad news to patients. *Baylor University Medical Center Proceedings, 29*(1), 101–102.

Mullan, R. (2017). *Mad to be Normal* (2nd ed.). London: Free Association Books

NHS. (2019). Euthanasia and physician-assisted suicide. https://www.nhs.uk/conditions/euthanasia-and-assisted-suicide/.

Nikookar, H. R., & Sooteh, S. H. J. (2014). Euthanasia: an Islamic ethical perspective. *European Scientific Journal*.

Novack, D. H., Plumer, R., Smith, R. L., Ochitill, H., Morrow, G. R., & Bennett, J. M. (1979). Changes in physicians' attitudes toward telling the cancer patient. *Journal of the American Medical Association, 241,* 897–900.

Oken, D. (1961). What to tell cancer patients: A study of medical attitudes. *Journal of the American Medical Association, 175,* 1120–1128.

Pence, G. (2016). *Medical ethics: Accounts of groundbreaking cases* (8th ed.). New York: McGraw Hill.

Rabow, M. W., & Mcphee, S. J. (1999). Beyond breaking bad news: How to help patients who suffer. *Western Journal of Medicine, 171*(4), 260.

Rao, R. N., Roy, S. K., & Tatiya, H. S. (2018). Dying with dignity: Physician assisted suicide in India: A critical review of legal facts. *Medico-Legal Update, 18*(1).

Ravitsky, V., & Prawer, M. (2008). The Dying Patient Law, 2005. *Jewish Medical Ethics and Halacha, 6*(2), 13–29.

Rothman, D. J. (1982). *Strangers at the bedside: A history of how law and bioethics transformed medical decision making*. New York: Basic.

Schultz, J. J. (2013). The Doctor's dilemma: The utilitarian medical ethics of Nazi physician Karl Brandt. *University of Toronto Medical Journal, 90*(4), 176–180.

Searight, H. R., & Barbarash, R. A. (1994). Informed consent: Clinical and legal issues in family practice. *Family Medicine, 26*(4), 244–249.

Searight, H. R., & Meredith, T. (2019). Physician deception and telling the truth about medical "Bad News": History, ethical perspectives, and cultural issues. In *The Palgrave Handbook of Deceptive Communication* (pp. 647–672). New York: Palgrave Macmillan.

Simon-Lorda, P., Tamayo-Velázquez, M. I., & Barrio-Cantalejo I. M. (2008). Advance directives in Spain. Perspectives from a medical bioethicist approach. *Bioethics, 22*(6), 346–354.

Sturman, R. L. (2003). *Six lives in Jerusalem: End-of-life decisions in Jerusalem: Cultural, medical, ethical and legal considerations.* New York: Springer.

Syed, A. A., Almas, A., Naeem, Q., Malik, U. F., & Muhammad, T. (2017). Barriers and perceptions regarding code status discussion with families of critically ill patients in a tertiary care hospital of a developing country: A cross-sectional study. *Palliative Medicine, 31*(2), 147–157.

Taupitz, J. (2013). Patient's autonomy according to German law. In S. Negri, J. Taupitz, A. Salkic, & A. Zwick (Eds.), *Advance care decision making in Germany and Italy* (pp. 111–129). Berlin: Springer.

Zwick, A. (2013). The German law on euthanasia: the legal basics and the actual debate. In S. Negri, J. Taupitz, A. Salkic, & A. Zwick (Eds.), *Advance Care Decision Making in Germany and Italy* (pp. 151–185). Berlin: Springer.

Chapter 4
Whether and How to Inform Patients of "Bad News," Family Dynamics at the End of Life

4.1 Whether and How to Tell: Physician—Patient Communication

Telling patients the truth about a life-threatening condition is less common outside of the United States. There is evidence that direct disclosure of medical "bad news" is more likely to occur in Northern, rather than Southern, Europe and probably more common in Europe overall than in many Asian countries (Blank, 2011; Searight & Gafford, 2005). Even in the United States, family requests for nondisclosure of serious medical conditions to patients are relatively common (McCabe, Wood, & Goldberg, 2010; Wang, Peng, Guo, & Su, 2013). Historically, physician disclosure of a cancer diagnosis was with vague terms such as "mass" or "growth" (Oken, 1961), most research suggests that contemporary U.S. physicians have become more direct with the patients.

Currently in the U.S., while diagnoses and treatment options are commonly disclosed, physicians are disinclined to make specific prognoses and often hedge prognostic information with qualifiers (Zier, Burack, Micco, et al., 2008). Research suggests that patients do not expect physicians to be precise in predicting patient outcomes. Patients recognize that the physician is making judgments in a realm of uncertainty and do not appear to judge physicians harshly for inaccurate prognostication when done in good faith and conveyed with sensitivity (Zier et al., 2009). Even when prognoses are ambiguous, a physician's honest, yet sensitive, description of likely outcomes prepares family members for the very real possibility of a loved one's death (Zier et al., 2009). Prognosticating the realistic death of a family member is likely to be associated with better psychological adjustment. There are suggestions that sudden, as opposed to anticipated, deaths are associated with greater levels of psychiatric disturbance (Zier et al., 2008).

H. R. Searight, *Ethical Challenges in Multi-Cultural Patient Care*,
SpringerBriefs in Ethics, https://doi.org/10.1007/978-3-030-23544-4_4

4.2 Cross Cultural Encounters in the U.S. and Canada: Interpreter Mediated Communication

Candib (2002) describes a case in which a U.S. physician, seeing a colleague's hospitalized Russian immigrant patient for the first time, directly disclosed a pancreatic cancer diagnosis. The patient's usual doctor was a primary care physician of Russian background who had instructed the hospital team not to tell the patient of his diagnosis. The patient's primary physician indicated that she would contact the family and they would make the decisions about the patient's care. This approach to informing the patent is still relatively common in Eastern Europe (Searight & Gafford, 2005). However, the hospital physician covering for his colleague, strongly believed that patients should be informed of their medical condition and would appreciate the information. The physician contacted the ATT language line and used their interpreter services to tell the patient of his diagnosis. The patient became angry and told the physician that he did not want to talk further and accused the doctor of taking away his hope (Candib, 2002).

With approximately 12–15% of the US population speaking a language other than English at home with a comparable figure of 6–8% for the United Kingdom (Searight, 2017, 2019), the likelihood of having a health care encounter mediated by a foreign language interpreter is relatively high. Patients from cultures in which it is common not to disclose serious medical information to patients—particularly older parents—may wish to have a family member serve as the interpreter. While this may seem intuitively helpful and is common, family members are not considered to be appropriate for filling this role in health care settings (Searight, 2017). Having a family member serve as an interpreter is explicitly forbidden by the Affordable Care Act (Searight, 2017). There are a number of issues that arise when family members serve in this role. A common problem is that family members may be bilingual but their skills in the second language may not extend to specialized areas such as health care. As a result, family members as well as untrained, ad hoc, interpreters may unintentionally misinterpret or omit clinically relevant information (Searight, 2017).

As has been noted previously, in Western countries it is legally required that the patient give informed consent to medical treatment. However, to give meaningful informed consent, the patient must be aware of their condition. There have been multiple situations reported in which family members were not willing to interpret physician provided information about cancer or a life threatening diagnosis. In cultures in which there are strong prohibitions about talking about death, such as in some Native American communities, family interpreters and some nonfamily interpreters may refuse to interpret health care professionals' verbal information regarding mortality and treatment options at the end of life (Ho, 2008). Another issue in cultures in which there is significant deference shown to elders is that having adult children interpret for a parent or grandparent does not respect traditional hierarchies (Searight & Searight, 2009).

Professional interpreters report significant moral dilemmas when physicians communicate directly to the patient about their diagnosis and prognosis and expect the interpreter to convey the information verbatim. This directive creates distress for the interpreter when they are from a culture in which this information is not typically explicitly conveyed to patients (Hordyk, MacDonald, & Brassard, 2017). This situation creates a genuine conflict for the interpreter between their professional health care role and their commitment to their culture.

Among the Inuit in northern Québec and Nunavut, conveying serious health information has traditionally been the task of community elders. These elders and local leaders are viewed as interpersonally wise and communicate information in ways that reflect sensitivity to the patient and family members' emotional well-being. For interpreters working in medical clinics in these regions in which tribal elders held this important role, interpreters often felt that that they were not the appropriate person to convey this information (Hordyk et al., 2017).

Research on Inuit health care has noted that in hospital settings in which interpreters were less likely to have personal relationships with patients and their extended families, a direct communication approach was more likely. However, when conveying news to local community members, interpreters were less direct. Hordyk and colleagues (2017) found that in these situations, interpreters were less likely to convey a specific prognosis but instead, tell the patient that only God would know when death would occur. Interpreters also indicated that they would add their own messages of hope to the physician's prognosis by saying things like "…you can lengthen the time, you can still have a livable, functional life even though you are given that time… Don't have to be stuck in bed." (Hordyk et al., 2017). Trying to maintain the patient's hope while doing their job as a healthcare professional, was often described as a lonely and distressing experience for these indigenous interpreters (Hordyk et al., 2017).

Similar conflicts have been described with interpreters working with Korean American families. Adult children, protective of their mother or father, would specifically tell the interpreter not to interpret diagnostic and prognostic information. This sets up an ethical dilemma for the interpreter who, professionally, has a responsibility to interpret everything the physician and patient say during the encounter. In a case involving a Vietnamese man with end-stage liver disease, a medical assistant in the physician's office served as the interpreter (Candib, 2002). At each visit, the interpreter explained to the physician that it was not appropriate or helpful to tell the patient of their diagnosis. The interpreter told the health care team that if she was mandated to translate the patient's diagnosis and prognosis, she would not do it. According to Candib (2002), the patient died having never been directly told of their medical condition.

4.2.1   Culture and Communication at the End of Life

In some European countries, particularly in the Mediterranean region, as well as in former East Bloc countries, there has often been a long standing practice of nondisclosure among physicians. Among a sample of Italians in Italy, fewer than half indicated that patients should always be informed of their diagnosis (Grasi et al., 2000; Surbone, Ritossa, & Spagnolo, 2004). However, only one out of four physicians indicated that they typically disclose this type of information to patients in their own practice. Among the sample of 675 doctors, nearly 1/3 indicated that patients did not want to know the truth about a serious medical condition. Surgeons were more likely to disclose and general practitioners were least likely with older physicians less likely to disclose the diagnosis than younger physicians (Grasi et al., 2000). While available information is somewhat dated, there does appear to be a greater likelihood that information is withheld from patients in southern Europe including Spain and Greece (Georgaki, Kalaidopoulou, Liarmakopoulos, & Mystakidou, 2002) as well as Italy, compared with northern Europe.

The continued diversity of attitudes and practices to informing patients, as well as changes in perspectives on this issue by age, are illustrated by a study conducted in Poland. Among a large group of over 200 physicians, 40% indicated that patients should be informed directly of an incurable condition with 59% of medical students agreeing to that position. Among practicing Polish physicians in this sample, 38% indicated that partially informing patients was appropriate with a comparable figure of 29% for medical students. Finally, 19% of physicians indicated that disclosure depended upon the patient's mental condition with 11% of medical students also endorsing this position (Lepper, Majkowicz, & Forycka, 2013).

However, it is reported that in Germany, where medical care is covered by national insurance, disclosure of diagnosis to the patient is required by law. Germany has also been mandating advance directives (Chattopadhyay & Simon, 2008). The availability of treatment as well as palliative care and its relationship to physician disclosures has not been well examined. It does seem logical, however that in countries in which life-prolonging measures and up-to-date palliative care are unavailable, there might be reluctance to disclose this information to patients or their families.

China has historically been a country in which patients were not informed of a life-threatening condition. While Chinese physicians have been moving more towards a Western model of informed consent, discretion around disclosure is still permitted (Wang et al., 2013). A law that went into effect in 2010 in China (Law of Medical Practitioners) holds that when disclosing diagnostic and treatment information, healthcare professionals have a responsibility to avoid actions that might adversely impact patients. Under Chinese tort law, the physician is held legally liable for any patient harm resulting from the disclosure (Wang et al., 2013).

Other studies conducted in China have found somewhat contradictory preferences. Both Chinese nurses and physicians generally report that the patient should not be informed of serious medical illness (Nie, 2013). At the same time, when nurses are asked to indicate what they would want if they were in the patient's situation they indicated that they would like to know their own diagnosis. At the same time however, nurses indicated they were reluctant to share this information with patients. Nearly 75% of this large sample of nurses indicated that they would deliberately withhold a cancer diagnosis from a patient. However, 90% of these nurses indicated that if they were in the role of patient, they would like this information (Nie, 2013). This particular sample of Chinese nurses was even reluctant to inform family members about a loved one's terminal condition-only 2.5% indicated they would inform family members immediately with 69% stating that they would eventually inform the family and 28% stating that they would never inform the patient's relatives of a terminal diagnosis (Nie, 2013). Comparable findings emerge in a study of oncologists in China with 60% indicating that terminally ill patients should not be informed of their condition (Jiang, Liu, Li, Huang, et al., 2007).

In the Middle East, there is still evidence of a pattern of withholding information from patients and instead, discussing diagnostic and prognostic information with close relatives. For example, in Saudi Arabia, 75% of physicians reported discussing information with family members instead of patients and in Kuwait, close to 80% of physicians indicated that, at the family's request, the physician would withhold diagnostic information from the patient (Sarafis, Tsounis, Malliarou, & Lahana, 2014). In Iran, close to half of terminally ill cancer patients were unaware of their diagnosis (Shahidi, 2010).

Illustrating the shift in both medical education as well as clinical practice, a survey of physicians in a specialized cancer center in Saudi Arabia indicated that over 80% reported being comfortable with disclosing "bad news" (Alshammary et al., 2017). While 86% indicated that they believed that patients should be directly informed of cancer, only 30% indicated they would do so if it was against the family's wishes. Also, 61% indicated that they would only give details to patients about the cancer condition if specifically asked. Again, indicating changes in both attitude in the medical community as well as in medical education, 70% of the sample reported receiving specific training in disclosure of medical bad news (Alshammary et al., 2017).

4.3 Family Dynamics

4.3.1 The Individual and Their Family

The family is the primary institution that mediates between the larger society, culture, and the individual. Social norms, religion, and patterns of relating to others are all lessons learned through family life (McGoldrick, Giordano, & Garcia-Preto, 2005).

The author has taught university students in China as well as Chinese nursing and medical students. During my first experience teaching in China, I was teaching developmental psychology with a focus on emerging adulthood. I asked the class of about 30 female students, "What makes someone an adult?" Typical responses from American students are that they are self-supporting financially and finished with their education. I pointed out to the Chinese students that in the United States having a child often was an indicator that one was an adult. The class seemed confused and I went on to elaborate that 30–40% of recent births in the United States were to single women. The class seemed genuinely astounded—it was almost as if they were saying to me "How can you have a child and not be married?"

It was useful for me to have this experience before teaching a medical ethics course to Chinese nursing and medical students. Rather than opening the class with a discussion of ethical theories, I spent the first few class periods describing American families. The students were very surprised about common family patterns in the United States such as cohabitation between unmarried partners, single women having children, as well as the high divorce and remarriage rate. I also presented them with data about the number of relationships engaged in by young adults before marriage. In the U.S., among those 20–24 years old, slightly more than 35% of females and almost 40% of males report having more than five sexual partners (Haderxhanaj, Leichliter, Aral, & Chesson, 2014). Among undergraduate women in China, only 18% reported having had sexual intercourse with approximately 5% reporting more than one sexual partner (Yan et al., 2009). This difference in perspectives about intimate relationships is part of a larger cultural emphasis on individualism in the United States and relationships often having secondary importance.

In Confucian and Islamic medical ethics, the family is the locus of important decisions about its members. Additionally, individual actions directly reflect upon the family and vice versa. For example, in many traditional societies, a young woman who has sex outside of marriage not only disgraces herself but her entire family. Similarly, if a family member has a psychiatric illness, the individual's mental health condition is often kept secret by the family since public knowledge would make young adult family members less desirable as marriage partners. In these cultures, marriage is not seen as the uniting of two individuals but as the merging of two extended families.

However, in the U.S., the family emphasis, while present, is modified by individualism. Both Cherlin (2009) and Cherry (2015) highlight the contradictions in the United States between the view that families are desirable and should be preserved and the emphasis on personal identity and self-development. Cherry (2015), in reviewing family demographics in the U.S. and Western Europe, argues that the influence of the family is in decline. The importance of autonomy and informed consent as an individual decision pushes the family to the sidelines: "The burden of proof is placed on the family to demonstrate that it acts with legitimate authority…" (Cherry, 2015, p. 43).

Cherry (2015) reviews a good deal of research indicating that the family with its long history of authority over individuals and as a valued social entity unto itself, is currently being challenged. Families demonstrate significant benefits both for children as well as for spouses. When taken as a group, children from divorced or single parent households are likely to exhibit more mental health issues, alcohol use and more likely to have their own marriages end in divorce. Females raised in single parent homes are much more likely to have a child while in high school. Additionally, married couples tend to have better mental and physical health (Cherry, 2015).

However, particularly in Western societies, the traditional family is under siege. In the U.S., approximately 30% of first births occur to unmarried women. In Sweden, more than half of first births occur to non-married, cohabiting couples (Cherlin, 2009).

Cherlin (2009) notes that for Americans, there is a tension between marriage and family and commitment to individual development. Marriage, divorce, remarriage, and marital cohabitation are relatively common occurrences in Western countries.

Americans "partner, partner, and [they] partner faster than residents of any other Western country" (Cherlin, 2009). By age 40, it is estimated that over 80% of American women will have been married. However, five years into their first marriage, 20% of Americans had separated or divorced. Cohabiting relationships are even more fragile with over half of them ending within five years. Among children in the United States, 40% will see a marital or parental cohabitation breakup by the time they are 15 years of age. In the U.S. up to half of all children who experience the end of their parents' marriage, experience a new partner with their parent within three years. Growing up in alternative family structures such as a cohabiting or remarried family is associated with greater risk of abuse as well as adolescents who are more likely to run away from home.

With this history of instability, it is certainly understandable that commitment to individual family members and particularly children would be diminished. Additionally, from a child's perspective, the array of adult parent like figures in their lives suggest that they will be living with parent like figures who are transient and may not have the child's best interest at heart.

By age 40, 80% of Hispanic women, 75% of Native American women and 90% of Whites, but fewer than two thirds of African-American women, are married. This pattern of decline in marriage among African-American women began in the 1960s (Raley, Sweeney, & Wondra, 2015) Hispanic and African Americans are more likely to end relationships with permanent separation versus divorce (Amato, 2010). In the absence of advance directives, this pattern of entering into a new relationship without legally terminating a previous one, may lead to conflicts between a current partner and an ex-spouse from whom the patient never formally divorced. Particularly when an estranged spouse is the patient's life insurance beneficiary, questions are raised when, in most states, the legal decision-maker (the estranged, but still legal husband or wife) is discouraging aggressive treatment or is seeking termination of life support.

In the U.S., particularly among those of Northern European background, marriages are more likely to be viewed as contracts. If one party does not hold up their end of the bargain, dissolution of the contract is understandable. This contractual model holds for relationships such as marriage. One of the best examples of this is the pre-nuptial agreement in which a future spouse protects their individual economic assets acquired prior to marriage. A frequent legal conflict occurring at the death of a parent and spouse is that between the deceased's children by a previous marriage and the deceased's most recent spouse. While again, probably not the optimal venue for resolving family issues, conflicts of this type often wind up in the courts. With a legal system based in individualism, court rulings reduce these collective conflicts to questions that are in favor or against an individual.

With families demonstrating patterns of instability and limited commitment to its members, the emphasis in U.S. medical ethics and law on individually based decision-making is understandable. Even in situations involving married couples, medical information cannot be directly shared. Each partner, despite their married status, maintains separate and private medical information. Indeed, the recently enacted

U.S. Health Insurance Portability and Accountability Act has been characterized as assuming that people want to keep information from their families (Cherry, 2015).

Cherry (2015) notes that in the West, the family is being further eroded by the power given to minors. He notes that, in the health care system, children are becoming "… self-possessed moral agents who undertake their own moral decision-making as soon as possible and as far as feasible" (Cherry, 2015, p. 53). Research ethics involving minors have been moving progressively in the direction of children having independent decisional authority, apart from their parents, to participate or not participate in healthcare studies. Additionally, many states such as California may not require parental consent for adolescents to obtain birth control or terminate a pregnancy (Cherry, 2015).

In 2014, Belgium approved an extension to their existing euthanasia law to permit physician-assisted assisted death for children. The clinical implementation of PAS for children is reportedly confined to cases in which children are "experiencing constant and unbearable suffering." While parental input is sought, in order for PAS to be implemented, the pediatric patient must demonstrate "the capacity for discernment" and recognize that choosing the intervention will end their life. The Belgian law highlights the role of the individual child as an independent decision-maker apart from their parents (Siegel et al., 2014). Children, in Belgium, have been authorized to independently seek physician assisted death. It is questionable whether the individual focus, particularly when it comes to minors, is actually in the individual child's best interests.

Cherry (2015) concludes that "…pressure has been brought to bear on parents and families through law and institutional policy, focusing among other concerns on separating children from the sphere of parental and familial authority as well as spouses from each other's authority" (p. 55). As has been noted at several points, in the U.S., in particular, conflicts around healthcare that fundamentally involve family relationships, are being addressed by the court. The Western legal system is individually focused and does not recognize families as coherent units for decision-making. Judges themselves have said that the courts are probably not the best place to resolve many healthcare issues but in a culture focusing on individual rights, the involvement of the legal system is understandable.

An example of how the legal system becomes involved in family matters is the Terri Schiavo case. In February, 1990, Ms. Schiavo, at age 26, experienced cardiac arrest at her home. She was resuscitated but was left comatose. After approximately two months, her diagnosis was described as a persistent vegetative state. For three years, her husband took Ms. Schiavo for experimental treatments and provided care himself. Speech and physical therapy was also attempted. In 1993, he asked that Terri's medical status change to Do Not Resuscitate.

In 1998, Mr. Schiavo asked the courts to remove Terri's feeding tube. Again, Ms. Schiavo did not have any written documentation as to her wishes. Terri's parents requested that life support including feeding, be continued. During the legal process, several pieces of uncorroborated evidence were cited by Terri's parents as evidence that Mr. Schiavo no longer represented their daughter's interests. Michael Schiavo had previously lived in the household with Terri's parents but moved out and eventually had a live-in female paramour. The parents' efforts to discredit Michael Schiavo included insinuations that he had physically abused Terri and that his actions even contributed to her comatose state at hospital admission. After several appeals, a county judge ordered the removal of the feeding tube. In all, there were a total of 14 appeals in the Florida courts. All the appeals to the federal courts upheld the decision to remove the tube. Ms. Schiavo died in late March 2005—15 years after she was placed on life support. The Schiavo case is an example of how when there is disagreement at the end-of-life in the U.S., the legal system mediates the final decision. The degree of acrimony and public airing of family differences were also features of this highly publicized case (Pence, 2016).

4.3.2   The Individual as Surrogate Decision-Maker

As noted above, the family is the primary transmitter of cultural and religious norms. Families, however, have their own internal dynamics regarding the locus of authority for important decisions, the degree of duty or obligation that members have towards one another, the relative importance of rules, consistency and structure, and the significance of tradition (Searight, 1997). In the face of a crisis, such as with a terminally ill parent, these patterns may be challenged or they may be taken to extremes such as when the oldest adult son tells his siblings that he, alone, will receive the information about their father's medical status and he, alone, will make all needed medical decisions.

In decision-making regarding a seriously ill family member, who may be demonstrating compromised cognitive functioning or who is non-responsive, U.S. hospitals both implicitly and explicitly continue the individualistic focus by attempting to establish a specific family member as the patient's surrogate. This pressure to immediately establish a family spokesperson may conflict with cultural norms that emphasize a process of collective consensus. As is often the case in Hispanic families, all members of the family may need to be convened before a decision can be communicated. The collective decision, may be ultimately communicated by a distinguished elder who moderates the discussion and speaks to healthcare professionals on behalf of the family.

In acute hospitals in the U.S., healthcare professionals prefer and actively try to install a single individual as the family spokesperson. In a qualitative study of end-of-life decision-making in the intensive care unit, Quinn, Schmitt, Baggs and colleagues (2012) developed a typology of how an individual came to be designated as the conduit of information for family members. The first role, the primary caregiver, was typically a family member who was involved with the patient's care before the hospital admission. Often, the eventual hospitalization felt like a "defeat" for the primary caregiver who had ultimately been unable to manage the patient successfully at home.

In other families, a primary decision-maker had been formally established through a durable power of attorney (DPOA). In one family, one of the patient's daughters was the legally indicated healthcare proxy. When it was required that a do not resuscitate order sheet be signed, the officially designated proxy signed first but then her siblings indicated that they all wanted to sign the DNR document. As one of her sisters said "We all want to sign, we don't want only [proxy] to feel like she's the one who did this" (Quinn et al., 2012, p. 5).

As mentioned above, clinical staff encouraged establishing a single-family spokesperson so that information can be transmitted efficiently (particularly important in large families) and to designate specifically who would be responsible for communicating any decisions that needed to be made. When this designation did not occur, as might be the case in families from collectivist cultures, the ICU staff often was frustrated. However, hospital staff also behaved in ways that conflicted with the designation of a specific spokesperson. For example if another family member happened to be visiting, they might become the temporary spokesperson (Quinn et al., 2012).

Another role was the "out-of-towner" (Quinn et al., 2012). The out-of-towner typically was a family member who had not been involved in ongoing care of the patient and often had not been present for the early stages of hospitalization. In some families, decisions about treatment were delayed until the out-of-towner arrived. However, the out-of-towner's arrival could trigger increased conflict. Having observed these dynamics as a hospital psychologist and a member of the institution's ethics committee, it is likely that the out-of-towner's perceived disruptiveness may stem from several motivations. Molloy and colleagues (1991) have described this pattern as "The Daughter from California Syndrome." An adult child who lives far away may experience guilt at not having been involved with their aging parent. To address their discomfort, they may seek aggressive treatment for their parent as a way of addressing the implication that they have been negligent (Molloy, Clarnette, Braun, Eisemann, & Sneiderman, 1991). Additionally, the out-of-towner often can only be present for a limited period of time and has to return their geographically distant home—often within several days. When a parent appears to be nearing death, the out-of-towner may want to expedite the process so they do not need to make a return trip for a funeral. The out-of-towner's motivation may also lead them to push for hospice care before other family member's have accepted the inevitably of the patient's relatively imminent death.

Another role in the U.S. ICU was "the patient's wishes expert." This family member was typically an adult child who was confident that they knew what the patient would want in terms of end-of-life care (Quinn et al., 2012). However, the "expert" may not have current knowledge of the patient's wishes and/or may interpret "what mom would want" differently from a sibling. As is often the case, if end of life issues are not often openly discussed and the time between discussion with a family member and serious illness is of some years' duration, the patient may have disclosed revised wishes to another family member. Often, two adult siblings might have very different interpretations about the patient's wishes.

Consistent with the concept of filial piety, there were often protectors—adult children who wanted to shield a vulnerable family member from having to make end-of-life decisions (Quinn et al., 2012). On rare occasions, the protector role included successfully convincing the health care team—at least temporarily—to withhold prognostic and treatment information from competent patients. While physicians would initially go to a competent patient for informed consent for treatment, it is assumed that the patient does have relevant information. If relevant information is being withheld, the patient cannot make a truly informed decision.

Finally, some families included a healthcare expert—a physician or nurse whose clinical expertise was respected by other family members. While family members with medical background could be helpful in terms of explaining procedures and outcomes of interventions such as intubation, at times they disagreed with the health care professionals who were actually taking care of their relative, Because the patient was a close family member such as an aging parent, emotion and loyalty often colored objective medical judgment (Quinn et al., 2012).

ICU staff observed that often there was not a single spokesperson but instead, a "spokes-group" that emerged when multiple family members wanted to actively participate in decision-making. These multiple spokespersons were particularly common when serious issues such as withdrawal of life support arose. As has been noted, this collective process has been described as particularly common among Hispanic communities s in the United States.

4.3.3 Cross-Cultural Perspectives on Families at the End of Life

Including all close family members in the process of decision-making for a loved one is often incongruent with the individually-focused medical and legal system found in countries such as the United States. For example, a variation on the "out-of-towner" theme may occur in non-Western and Hispanic cultures but is not typically associated with social disruption and interpersonal conflict. While not formally established as law, in Korea and Taiwan, a standard practice is that physicians communicate medical bad news to the oldest member or head of the family. However, in these Asian countries, it is also extremely important that all family be members be present at the time of their loved one's death (Cheung et al., 2015).

Among African-American families, while there was often agreement about maintaining the dignity of the terminally ill family member, issues such as denial of the significance of the illness and differences in religious views often were the sources of internal family conflict (Johnson, Hayden, True, et al., 2016).

While not typically described as highly collectivist in their orientation, the U.S. African-American community typically places a very strong value on family. At the same time, consensus may be more difficult to achieve when it comes to an ailing family member. In a study of palliative care providers, African-American families were described by physicians as supportive yet, complex (Rhodes, Batchelor, Lee, & Halm, 2015). In the health care professional's opinion, the patient, themselves, was often ready to move towards accepting palliative care but family members persisted in seeking aggressive treatment. It was noted that the patient, themselves, may feel pressured by family members' agendas to "keep fighting" and may continue with unsuccessful treatment to please an adult son or daughter (Rhodes et al., 2015).

This complexity was evident in Johnson et al.'s (2016) study of end of life care among African Americans. These investigators described how family members were often very engaged in issues surrounding end of life care (Johnson et al., 2016), but, at the same time, there was often pronounced dissension within the family: "In my family, they just cannot make decisions." (Johnson et al., 2016, p. 145). Another respondent mentioned regrettable conflict: "… the last thing that resonated with me from my father was I know it was very important to him that his family be together, but we were not all at the same place." (Johnson et al., 2016, p. 145).

In dealing with patients from Asian and some Hispanic cultures, the Western physician may experience a desire to free the individual patient from the perceived tyranny of familial control (Ho, 2008). While it may be perceived that the family is exerting undue influence on the patient, family members may view themselves as protective—particularly in the face of a healthcare professional who may not understand their family's culture.

As Western views of patient autonomy become more common in Africa and Asia, there has been increased interest in advance directives in these regions. However, the introduction of advance directives may result in conflicts between traditional collectivist versus newly-introduced individualist approaches to decision-making. Foo and colleagues (2012) note that when the patient does not exhibit capacity for independent decision-making, the family as a unit becomes the decision-maker. However, durable powers of attorney are only available for one person. If these are enacted in collectivist cultures, family members may not accept the "official" status of the person designated as the power of attorney. Complicating matters further is that surrogates often want a more aggressive level of care than the patient, themselves (Suhl, Simons, Reedy, & Garrick, 1994).

One of the key conflicts that emerges in collectivist societies is that the patient may also experience a responsibility to protect their family. Reflecting a communitarian approach, Candib (2002) presents the case of Mr. Doe, a 42-year-old Vietnamese man diagnosed with metastatic squamous cell carcinoma. After being told of the diagnosis and the poor prognosis, Mr. Doe requested that the physician not inform his family. "They have enough on their minds already they do not need to worry about this." While the physician originally thought that the patient was "in denial", over the next several months, as he deteriorated, the patient did inform his wife and daughters. He went back for to Vietnam for a visit and returned and died soon after.

This theme of not wanting to burden family members also emerged in a study of Korean Americans. It was found that interviewees who were concerned about being a burden to family members were more likely to have engaged in end-of-life discussions. In addition, older Korean-Americans who were particularly concerned about being a burden on family were less likely to accept life sustaining treatments and also less likely to complete advance directives (Duke, Thompson, & Hastie, 2007; Ko, Roh, & Higgins, 2013).

The web of family duties is complex. In Frank et al.'s (1998) interview study of Korean Americans, their informant, Mrs. Kim, noted that dying at home is the preference in Korea. However, one benefit she observed in the American custom of dying in a hospital is that it reduces the family's responsibility. At the same time, however, Mrs. Kim indicated that it is the family's responsibility to do anything they can to prevent their loved one's death. A contradiction that emerged was that Mrs. Kim indicated that she would not be controlling her treatment if she became terminally ill. When talking about herself, Mrs. Kim indicated she would not want to prolong futile treatment. However, in following her preference for family-based decision-making, she saw her family as ethically obligated to request that everything should be done to maintain her life even in the face of medical futility (Frank et al., 1998).

Among the Roma who have settled throughout Europe including in Romania, Hungary, Germany, France and Spain, there is a very strong commitment to family with multiple generations living in a single household and considerable importance devoted to family rituals such as weddings and births. Finally, there is a strong value of respect for and care for the elderly. It is expected that adult children care for their parents and other older family members. Illness is not seen as an individual problem but as a phenomenon affecting the extended family as a whole (Peinado-Gorlat et al., 2015). The Roma's strong emphasis on extended family includes obligation to more distant relatives such as uncles and aunts, as well as nieces and nephews (Peinado-Gorlat et al., 2015). Women in the Roma culture are the caregivers for these elderly family members. For Roma women, this responsibility is an automatic duty and considered a priority.

The contradiction noted in Asian-American families was also reported by Roma women. When asked hypothetically about their own wishes at the end-of-life, there was a strong preference for not being maintained alive through technology. However, when it came to a family member, doing everything possible to maintain life was seen as a duty or obligation. Family members were described as suffering more at the end of a patient's life than the patient, themselves: "You're suffering, but those around you are suffering more. That's even worse." (Peinado-Gorlat et al., 2015).

The degree of collectivism among the Roma is illustrated by a condition called resignation syndrome. This condition, described among Roma immigrant children in Sweden, occurs when the family has been formally denied legal asylum by the Swedish government. Soon after the denial, children and adolescents enter a comatose-like state and may require tube feeding. Resignation syndrome resolves when the family successfully appeals and receives legal asylum (Aviv, 2017).

In many states within the U.S., when a patient has not specified a surrogate or does not possess a clear advance directive, there is a succession policy of persons who may decide on behalf of the patient. Even in states without formal legislation, this sequence is typically followed by physicians to determine who shall. make medical decisions for an incompetent patient who cannot represent their own desires. The typical "chain-of-command" for medical decision-making in order of priority is: —patient's spouse, followed by the patient's parents, followed by adult children, and other relatives in order of consanguinity. In clinical situations, if conflicts arise, they are most likely to occur between adult children. Among the patient's offspring, other than requiring that they be of age, there is no stated sequence for determining who should serve as surrogate decision-maker. When an aging parent does not have decisional capacity and there are multiple adult children, conflicts may arise without a clear legal framework for resolving them.

4.3.4   Hardwig: At the End-of-Life, Patients Are an Unfair Burden to Families

In U.S. bioethics, the major proponent of a family focus is the philosopher, John Hardwig. Hardwig has written a series of provocative and controversial articles such as "Is There a Duty to Die," in which he argues that the emotional and financial strain on the family coping with a terminally ill member may outweigh the value of extending the patient's life. An anthology of Hardwig's writings includes an afterword in which members of Hardwig's family share their reflections about the duty to die. Included is Hardwig's son who shares a story from the extended family regarding his great-grandfather (John Hardwig's grandfather). After suffering a significant heart attack, the great-grandfather was ordered by his physician to have an extensive period of bed rest. Because he was concerned about the imposition that he would be placing upon the family, and the fact that the family had little in the way of health insurance, he committed suicide. Hardwig implies that his great grandfather calculated that the money from his life insurance would be of greater benefit to his wife than his continued presence in a debilitated state.

With illustrative examples such as these, Hardwig argues that contemporary U.S. bioethics suffers from an individualistic fantasy. "Lives are separate and unconnected or that they could be so if we chose…. I …[am] …free morally to live my life however I please, choosing whatever life and death I prefer for myself" (Hardwig, 2000, p. 121). In reality, in end-of-life care, when the question arises "What is best for the patient?", the patient is focused upon as an autonomous individual without responsibilities to others. Hardwig (1990) argues that because of the impact that their decisions have on family members, patients may have a duty to refuse treatment that prolongs the dying process.

While not as relevant in countries with national, universal health coverage, Hardwig (1990) points out the significant financial strain the dying process places on families in the United States. While this burden may be mitigated somewhat by the Affordable Care Act, the fact that there is still 7–10% of the U.S. population that "falls through the cracks" with respect to insurance and that adequate levels of health insurance coverage are often accompanied by exorbitant premiums which many individuals cannot afford, Hardwig's financial analysis, while often raising discomfort, addresses realistic issues.

In the U.S., which does not have government-based universal health coverage and in which private insurance companies may charge substantial premiums and still limit the amount that they will pay for care, a loved one's serious illness can have significant financial impact on the family. Hardwig (2000) notes that among patients who had less than a 50% chance of living six months, 20% of them had family members who had quit their jobs or made significant economically-related lifestyle changes, close to 30% lost their savings and another 30% lost a significant amount of income (Hardwig, 2000) in caring for the ailing family member. While it is morally distressing to perform economic analyses on a loved one's survival, Hardwig (2000) makes the case that these additional six months of life are likely of less than optimal quality and, from a utilitarian perspective, may not be worth the fiscal burden placed on families. This is particularly noteworthy in the United States, where family members provide support for the vast majority of older individuals in ill health.

Hardwig (2000) goes as far as to generate a list of circumstances that "make it more likely that one has a duty to die" (p. 129). These circumstances include age—as one grows older, there is less of life to be sacrificed for others; "your loved ones have already made great contributions-perhaps even—sacrifices to make your life a good one." (Hardwig, p. 129).

Hardwig (2000) even addresses the issue of competence. He argues that if one has become incompetent through some type of dementia "the part of you that is loved will soon be gone or seriously compromised." If one sees that cognitive incapacity is likely, Hardwig believes that one may have the duty to end one's own life before the medical and legal systems become involved. He points out that the medical and legal systems are likely to prolong one's life even when it is burdensome to others. However, unlike cultures that are collectivist in their orientation, Hardwig does not describe families as having a shared cohesive culture. He sees that families are composed of individuals—each with their own beliefs, values, and goals. Hardwig's (2000) perspective is certainly logical and rational. However, human beings frequently are not so cognitively analytic. The idea of ending one's life early to benefit others creates a scene of spending ones' final days making a "pro" and "con" list.

What makes Hardwig's perspective different than the collectivism just described for Roma and Asian families? The families in these culturally distinct communities share a collective identity with accompanying core values. The families described by Hardwig are groups of individuals concerned with their own self-development. In Asian collectivism, there is an emphasis on shared family feeling. In contrast to Hardwig's view of colliding atoms, discussions of East Asian family-oriented ethics describe a complex web of relationships: "one is no longer an independent person but a member of the family, a unified ethical entity" (Cai, 2015, p. 191).

4.4 The Physician as Part of the Family

In countries in which Confucianism and Islam exert significant influence, physicians are still revered figures. While the issues described by Quinn et al. (2012), in the U.S. intensive care unit have an "us versus them" connotation with the medical team pitted against the patient's family, this separation of the health care professional from the patient's family reflects distinct cultural values.

In China and to some extent South Korea, the Confucian ethic places the physician within the family, itself. One of the virtues of Confucianism is that the physician, in whom the family places great trust, would treat their patients as if they were the doctor's own family members (Lee, 2015). The physician, to use Pellegrino's term, exhibits phronesis—wisdom—which engenders this trust. Loyalty is another key element of Confucianism and the healthcare professional demonstrates this: fidelity to the patient and their family. A recent study conducted in Hong Kong, with both patients and their families as well as healthcare professionals, found that joint decision-making by the family and healthcare team was ranked most highly. In discussing the possibility of advance directives, the majority again viewed discussion between the family, the patient, and the physician as the optimal way of drawing up an agreement of this type. The sequence for this approach to decision-making is essentially that the physician and other members of the healthcare team discuss the optimal treatment while including attention to the patient's preferences and present the plan to the family (Chan, Doris, Wong, et al., 2015).

Moazam (2000) describes how in Pakistan, it is common practice to protect patients by not disclosing a terminal condition. However, the physician, is also seen as a member of the family. They are treated with great respect—often addressed as *Sahib* ("Lord"). In Pakistan, there is a strong respect for authority and extended families typically live together even after marriage. One's identity is primarily through family membership. Moazam (2000) describes how she often experienced herself as a distinguished elder in the families of patients that she treated. It was not unusual for patients and their families to ask her "Doctor, *Sahib*, what you would do if you were in my place?" In an interaction with the father of a critically ill newborn, Moazam (2000), following her training in the West, began describing the infant patient's condition. She was interrupted by the father who indicated that he placed his faith in God and that Dr. Moazam represented the wisdom of God as well as the healing power of God on earth. He did not need any more information; the father trusted Dr. Moazam to do what was best for the infant and their family.

4.5 Conclusion

While disclosure of serious medical conditions as well as treatment options are essential parts of informed consent, this model of disclosure is not universally accepted. While the United States and other Western countries are undergoing profound changes that have left families in states of interpersonal instability and often with few individuals in whom they can consistently trust, outside of Western industrialized countries, which comprise only a minority of the world population, family cohesiveness is still common. It is not surprising that the development of advance directives and durable powers of attorney which permit the individual to continue to express preferences for care when they can no longer communicate their desires, have become more common in the United States than elsewhere. In a society in which the only one who can be counted upon is oneself, healthcare decisions centered on individual autonomy are certainly understandable. However, changes in the U.S. family also make it easier to see how, in many respects, Western countries are, presently an individualistic anomaly in their approach to end-of-life decision-making.

References

Amato, P. R. (2010). Research on divorce: Continuing trends and new developments. *Journal of Marriage and Family, 72*(3), 650–666.

Alshammary, S. A., Hamdan, A. B., Tamani, J. C., Alshuhil, A., Ratnapalan, S., & Alharbi, M. (2017). Breaking bad news among cancer physicians. *Journal of Health Specialties, 5*(2), 66.

Aviv, R. (2017, April 3). The trauma of facing deportation. *The New Yorker.*

Blank, R. H. (2011). End-of-life decision making across cultures. *The Journal of Law, Medicine & Ethics, 39*(2), 201–214.

Cai, Y. (2015). On family informed consent: On the legislation of organ donation in China. In R. Fan (Ed.), *Family-oriented informed consent: East Asian and American perspectives* (pp. 187–200). New York: Springer.

Candib, L. M. (2002). Truth telling and advance planning at the end of life: Problems with autonomy in a multicultural world. *Families, Systems, & Health, 20*(3), 213.

Chan, H. M., Doris, M. T., Wong, K. H., Lai, J. C. L., & Chui, C. K. (2015). End-of-life decision making in Hong Kong: the appeal of the shared decision making model. In *Family-Oriented Informed Consent* (pp. 149–167). Cham: Springer.

Chattopadhyay, S., & Simon, A. (2008). East meets West: Cross-cultural perspective in end-of-life decision making from Indian and German viewpoints. *Medicine, Health Care and Philosophy, 11*(2), 165–174.

Cheng, S. Y., Suh, S. Y., Morita, T., Oyama, Y., Chiu, T. Y., Koh, S. J., ... Tsuneto, S. (2015). A cross-cultural study on behaviors when death is approaching in East Asian countries: What are the physician-perceived common beliefs and practices? *Medicine, 94*(39), 1–5.

Cherlin, A. (2009). *Marriage, divorce, remarriage.* Cambridge, MA: Harvard University Press.

Cherry, M. J. (2015). Individually directed informed consent and the decline of the family in the West. In *Family-Oriented Informed Consent* (pp. 43–62). Cham: Springer.

Duke, G., Thompson, S., & Hastie, M. (2007). Factors influencing completion of advanced directives in hospitalized patients. *International Journal of Palliative Nursing, 13*(1), 39–43.

Foo, W. T., Zheng, Y., Yang, G. M., Kwee, A. K., & Krishna, L. K. R. (2012). Factors considered in end-of-life decision-making of healthcare professionals. *BMJ Supportive & Palliative Care, 2*(Suppl 1), A45–A46.

Frank, G., Blackhall, L. J., Michel, V., Murphy, S. T., Azen, S. P., & Park, K. (1998). A discourse of relationships in bioethics: Patient autonomy and end-of-life decision making among elderly Korean Americans. *Medical Anthropology Quarterly, 12*(4), 403–423.

Georgaki, S., Kalaidopoulou, O., Liarmakopoulos, I., & Mystakidou, K. (2002). Nurses' attitudes toward truthful communication with patients with cancer: A Greek study. *Cancer Nursing, 25*(6), 436–441.

Grassi, L., Giraldi, T., Messina, E. G., Magnani, K., Valle, E., & Cartei, G. (2000). Physicians' attitudes to and problems with truth-telling to cancer patients. *Supportive Care in Cancer, 8*(1), 40–45.

Haderxhanaj, L. T., Leichliter, J. S., Aral, S. O., & Chesson, H. W. (2014). Sex in a lifetime: Sexual behaviors in the United States by lifetime number of sex partners, 2006–2010. *Sexually Transmitted Diseases, 41*(6), 345–352.

Hardwig, J. (1990). What about the family? *Hastings Center Report, 20*(2), 5–10.

Hardwig, J. (2000). *Is there a duty to die?: And other essays in bioethics.* Routledge.

Ho, A. (2008). Using family members as interpreters in the clinical setting. *The Journal of Clinical Ethics, 19*(3), 223–233.

Hordyk, S. R., Macdonald, M. E., & Brassard, P. (2017). End-of-life care in Nunavik, Quebec: Inuit experiences, current realities, and ways forward. *Journal of Palliative Medicine, 20*(6), 647–655.

Jiang, Y., Liu, C., Li, J. Y., Huang, M. J., Yao, W. X., Zhang, R., … & Zhao, X. (2007). Different attitudes of Chinese patients and their families toward truth telling of different stages of cancer. *Psycho-Oncology: Journal of the Psychological, Social and Behavioral Dimensions of Cancer, 16*(10), 928–936.

Johnson, J., Hayden, T., True, J., Simkin, D., Colbert, L., Thompson, B., … Martin, L. (2016). The impact of faith beliefs on perceptions of end-of-life care and decision making among African American church members. *Journal of Palliative Medicine, 19*(2), 143–148.

Ko, E., Roh, S., & Higgins, D. (2013). Do older Korean immigrants engage in end-of-life communication? *Educational Gerontology, 39*(8), 613–622.

Lee, I. (2015). Filial duty as the moral foundation of caring for the elderly: its possibility and limitations. In *Family-Oriented Informed Consent* (pp. 137–147). Cham: Springer.

Leppert, W., Majkowicz, M., & Forycka, M. (2013). Attitudes of Polish physicians and medical students toward breaking bad news, euthanasia and morphine administration in cancer patients. *Journal of Cancer Education, 28*(4), 603–610.

McCabe, M. S., Wood, W. A., & Goldberg, R. M. (2010). When the family requests withholding the diagnosis: Who owns the truth? *Journal of Oncology Practice, 6*(2), 94–96.

McGoldrick, M., Giordano, J., & Garcia-Preto, N. (Eds.). (2005). *Ethnicity and family therapy.* New York: Guilford Press.

Moazam, F. (2000). Families, patients, and physicians in medical decision making: A Pakistani perspective. *Hastings Center Report, 30*(6), 28–37.

Molloy, D. W., Clarnette, R. M., Braun, E., Eisemann, M. R., & Sneiderman, B. (1991). Decision making in the incompetent elderly: "The daughter from California syndrome." *Journal of the American Geriatrics Society, 39*(4), 306–309.

Nie, J. B. (2013). *Medical ethics in China: A transcultural interpretation.* New York: Routledge.

Oken, D. (1961). What to tell cancer patients: A study of medical attitudes. *Journal of the American Medical Association, 175,* 1120–1128.

Peinado-Gorlat, P., Castro-Martínez, F. J., Arriba-Marcos, B., Melguizo-Jiménez, M., & Barrio-Cantalejo, I. (2015). Roma women's perspectives on end-of-life decisions. *Journal of Bioethical Inquiry, 12*(4), 687–698.

Pence, G. (2016). *Medical ethics: Accounts of groundbreaking cases* (8th ed.). New York: McGraw Hill.

Quinn, J. R., Schmitt, M., Baggs, J. G., Norton, S. A., Dombeck, M. T., & Sellers, C. R. (2012). Family members' informal roles in end-of-life decision making in adult intensive care units. *American Journal of Critical Care, 21*(1), 43–51.

Raley, R. K., Sweeney, M. M., & Wondra, D. (2015). The growing racial and ethnic divide in US marriage patterns. *The Future of children/Center for the Future of Children, the David and Lucile Packard Foundation, 25*(2), 89.

Rhodes, R. L., Batchelor, K., Lee, S. C., & Halm, E. A. (2015). Barriers to end-of-life care for African Americans from the providers' perspective: Opportunity for intervention development. *American Journal of Hospice and Palliative Medicine, 32*(2), 137–143.

Sarafis, P., Tsounis, A., Malliarou, M., & Lahana, E. (2014). Disclosing the truth: A dilemma between instilling hope and respecting patient autonomy in everyday clinical practice. *Global Journal of Health Science, 6*(2), 128.

Searight, H. R. (1997). *Family of origin therapy and cultural diversity.* Philadelphia: Taylor & Francis.

Searight, H. R. (2017). Clinical and ethical issues in working with a foreign language interpreter. *Journal of Health Service Psychology, 43,* 79–82.

Searight, H. R. (2019). *Conducting psychotherapy with an interpreter.* Oxford Handbook of Psychotherapy Ethics, in press.

Searight, H. R., & Gafford, J. (2005). "It's like playing with your Destiny": Bosnian immigrants' views of advance directives and end-of-life decision-making. *Journal of Immigrant Health, 7*(3), 195–203.

Searight, H. R., & Searight, B. K. (2009). Working with foreign language interpreters: Recommendations for psychological practice. *Professional Psychology: Research and Practice, 40*(5), 444.

Shahidi, J. (2010). Not telling the truth: Circumstances leading to concealment of diagnosis and prognosis from cancer patients. *European Journal of Cancer Care, 19*(5), 589–593.

Siegel, A. M., Sisti, D. A., & Caplan, A. L. (2014). Pediatric euthanasia in Belgium: Disturbing developments. *JAMA, 311*(19), 1963–1964.

Suhl, J., Simons, P., Reedy, T., & Garrick, T. (1994). Myth of substituted judgment: Surrogate decision making regarding life support is unreliable. *Archives of Internal Medicine, 154*(1), 90–96.

Surbone, A., Ritossa, C., & Spagnolo, A. G. (2004). Evolution of truth-telling attitudes and practices in Italy. *Critical Reviews in Oncology/Hematology, 52*(3), 165–172.

Wang, D. C., Peng, X., Guo, C. B., & Su, Y. J. (2013). When clinicians telling the truth is de facto discouraged, what is the family's attitude towards disclosing to a relative their cancer diagnosis? *Supportive Care in Cancer, 21*(4), 1089–1095.

Yan, H., Chen, W., Wu, H., Bi, Y., Zhang, M., Li, S., & Braun, K. L. (2009). Multiple sex partner behavior in female undergraduate students in China: A multi-campus survey. *BMC Public Health, 9*, 305.

Zier, L. S., Burack, J. H., Micco, G., Chipman, A. K., Frank, J. A., Luce, J. M., & White, D. B. (2008). Doubt and belief in physicians' ability to prognosticate during critical illness: the perspective of surrogate decision makers. *Critical care medicine, 36*(8), 2341. https://www.ncbi.nlm.nih.gov/pmc/articles/PMC2628287/pdf/nihms-85681.pdf.

Zier, L. S., Burack, J. H., Micco, G., Chipman, A. K., Frank, J. A., & White, D. B. (2009). Surrogate decision makers' responses to physicians' predictions of medical futility. *Chest, 136*(1), 110–117.

Chapter 5
Advance Directives, Do Not Resuscitate Orders, Hospice, Organ Transplantation and Physician Assisted Suicide

Advance care planning can take a number of forms including discussions with family members that are not formally documented, similar discussions with one's physician, and directives for level of care during hospitalization such as do not resuscitate (DNR) orders. Organ transplantation has, in the span of less than 50 years, moved from a rare, highly publicized experimental procedure to a relatively common surgical procedure for conditions such as kidney failure. A difficult issue with harvesting and transplanting organs such as the heart, is that the donor, while maintaining some physiological activity, is considered to be in an irreversible state of death. As was evident in the description of ethical dilemmas in Israel, death has not been consistently defined. In addition, there are significant cultural differences as well as international variability in the use of advance directives, hospice, do not resuscitate orders, and the acceptability of physician-assisted death.

5.1 Advance Directives

Advance Directives (ADs) have been mentioned throughout this book. In this section, these documents will be examined in detail. Advance directives are documents which indicate the patient's wishes regarding treatment when they are no longer able to convey their requests for care or in which, due to cognitive compromise, there is concern about a patient's judgment and decisional capacity. These documents, in part influenced by the attention given to the Quinlan and Cruzan cases, permit the patient to maintain autonomous decision-making regarding their medical care. A living will is a document that describes the level of treatment and level of life support desired. A living will may be used in conjunction with a Durable Power of Attorney which is a specifically named individual who represents the patient's decisions about treatment and who serves as proxy for the patient. In some European countries, both approaches are used simultaneously. The individual appointed durable power of attorney oversees the implementation of the living will to be certain that its terms are carried out as the patient intended.

© The Author(s), under exclusive license to Springer Nature Switzerland AG 2019
H. R. Searight, *Ethical Challenges in Multi-Cultural Patient Care*,
SpringerBriefs in Ethics, https://doi.org/10.1007/978-3-030-23544-4_5

While initially described in the 1970s, ADs became increasingly common in the U.S. with the passage of the Patient Self-Determination Act, which required that every health care institution receiving Federal funds (typically in the form of Medicaid reimbursement) provide information to all patients about ADs. As a part of the Affordable Care Act, the Centers for Medicare and Medicaid Payment are to provide reimbursement for end-of-life counseling. Politicians, who were opposed to this provision, were often opposed to the Affordable Care Act, in general, and referred to physicians having these discussions with patients as being part of "death panels." (Frankford, 2015). While the Affordable Care Act is undergoing revision, the original legislation established a policy that there should be no additional cost to Medicare patients if these discussions occur as part of annual checkups.

5.1.1   How Many People Have Advance Directives?

In the U.S., depending upon the setting and the population sampled, there is considerable variability in the prevalence of advance directives (ADs). It is estimated that slightly more than 1/3 of the general population has an advance directives (Yadav et al., 2017). A large-scale study examining the frequency of ADs, based on data from 2011 to 2016, concluded that 36.7% of the US population had completed an advance directive which included 29.3% who had a living will. Of note, among those with chronic illnesses, 38.2% had a completed advance directive compared with 32.7% of "healthy" adults (Yadav et al., 2017). A study conducted 15 years after the passage of the PSDA found that 18% of all patients hospitalized in medical and surgical units had an advance directive (Morrell, Brown, Qi, Drabiak, & Helft, 2008).

The prevalence of advance directives varies based on the age of patients as well as the geographic location of the setting in which they are hospitalized. It is estimated that about 70% of older adults in the community have an advance directive. While there is geographic variation in the presence of ADs, these regional differences become less pronounced when factors such as patient race and level of Medicare expenditure are included (Nicholas, Langa, Iwashyna, et al., 2011).

The need for formalized end-of-life decision guidance is supported by the types of issues commonly arising in hospitals. In a study of patients over 60 years of age who died in a hospital setting, slightly over 40% needed decisions to occur about their care (Silveira, Kim, & Langa, 2010). Characteristics of patients needing this assistance with decision-making were patients who: exhibited significant memory problems, were admitted from nursing homes, had experienced the death of a spouse and were diagnosed with cerebrovascular disease. It was estimated that of those clinical situations in which decision-making about the patient's care was an issue, 70% of the patients did not exhibit intact decisional capacity. Among those with living wills, only 2% requested aggressive treatment while 96% requested comfort care. Among this group of patients, living wills had been completed a median of 20 months before death (Silveira et al., 2010).

As the average lifespan lengthens and more people live well into the 80s, the likelihood of cognitive impairment increases as well. Therefore, from the perspective of future public health planning, advance directives are consistent with demographic trends. It has been reported that a written advance directive can be completed in a median time of 14 min in a physician's office visit. However, the range of time intervals for completing an AD varied considerably from −8 to 44 min. While both physicians and patients appear to see the value in advance directives, the percentage of completed directives remains fairly stagnant. Generally speaking, patients prefer that their physicians initiate these discussions and if they do so, patients appear to be willing to participate (Emanuel, Barry, Stoeckle, Ettelson, & Emanuel, 1991). However, despite being logically predisposed to advance directives, physicians often do seem to be reluctant to bring up the topic with patients. The outpatient setting in which patients are generally fairly healthy would seem to be the preferred context for completing ADs compared with an inpatient hospital admission precipitated by life-threatening illness.

5.1.2 Diversity and ADs in the United States

In the U.S., White European patients are most likely to have an AD. African-Americans as well as Hispanics are significantly less likely to use an AD (Johnson, Kuchibhatla, & Tulsky, 2008). In part, this pattern may stem from a spiritual perspective that death is something that is determined by God and that it is an affront to God's plan for the patient or the family to take over this process. Advance directives in general are used to limit life-sustaining care—African-American and Hispanic patients are more likely to desire ongoing care despite a poor prognosis (Blackhall, Murphy, Frank, Michel, & Azen, 1995; Searight & Gafford, 2005). Among immigrants to the United States, those with greater degrees of acculturation were more likely to have engaged in some type of advance care planning (Blackhall et al., 1995).

While religion and spirituality are likely to be significant influences on AD completion among some ethnic communities, the general population does not demonstrate an association between religiosity or involvement in faith-based activities with advance care planning. This global absence of an association is noteworthy since religion and spirituality often are significant issues at the end of life. One of the implied reasons for not participating in end-of-life discussions among those who view religion as an important part of their life is that issues of life and death are appropriately left to God. With respect to religious affiliation, participants reporting Catholic and evangelical affiliations were less likely than mainline Protestant affiliates to have advance directives even after adjusting for personal health values (Blackhall et al., 1995). Conservative Protestants were more likely to report that religious beliefs would influence their medical decisions and to endorse the view that the length of one's life is determined by God.

5.1.3 Prevalence of ADs Outside the United States

In Canada, Belgium, France, and much of Australia, advance directives have been formally codified into law (Beširević, 2010). While currently undergoing changes in end-of-life practices, collectivist cultures have, historically, had lower rates of AD completion. In a sample of 400 t Japanese citizens, over 80% wanted information about their diagnosis and treatment options. However, fewer than half of those surveyed wanted a formal, written, advance directive. The majority preferred to convey their wishes regarding treatment verbally to the physician or family members. Among residents of a palliative care institution in Korea, only 0.7% had an advance directive (Lee, 2015). As of 2015, ADs had not received legal recognition in China (Yang, 2015). While the documents have been approved for use in Taiwan in 2000, their use is restricted to terminally ill patients who have received the approval of two physicians (Yang, 2015).

In Hong Kong, ADs were also rare. In a study of health care providers, patients, and their families, advance directives were seen as representing the patient's wishes; however, family members and healthcare staff did not view these documents as providing guidance in any absolute sense. ADs were viewed as one piece of information to be used along with family and physician input. Nearly 2/3 of family members and close to half of patients themselves indicated that there were situations in which it was appropriate not to follow the patient's advance directive (Chan, Doris, Wong, et al., 2015).

Among the Roma of Europe, ADs were not seen as appropriate. In many respects these documents were seen as insensitive and almost disrespectful "You're going to read a paper on how to care for your mother? You care for her in her own way; you're not going to read a piece of paper" (Peinado-Gorlat, Castro-Martínez, Arriba-Marcos, Melguizo-Jiménez, & Barrio-Cantalejo, 2015).

Recent Bosnian immigrants to the U.S. also expressed discomfort with the emphasis on formally documenting these decisions: "Its s like you've already planned to live or die. We don't do this like you do in this country… It's too much; you only have one life" (Searight & Gafford, 2005, p. 301).

5.1.4 Limitations of ADs

Early research on the impact of ADs indicated that they often had little impact on limiting treatment (Perkins, 2007). There also did not appear to be much correspondence between patients' wishes as expressed through the advance directive and actual care received. While the assumption behind advance directives is that impaired decision-making ability is a common feature in older seriously ill patients, the actual use of ADs in with patients exhibiting impaired decisional capacity who are experiencing life-threatening illness with potential for treatment is unknown (Silveira et al., 2010).

One of the key conflicts with advance directives is that family members of patients with cognitive impairment or who are nonresponsive often want more aggressive treatment than the patient's living will specifies.

5.2 Do Not Resuscitate (DNR) Orders and Limiting Care

Do not resuscitate (DNR) orders permit (usually hospitalized) patients to indicate if they would want the health care team to provide cardiopulmonary resuscitation (CPR) in the event of a cardiac arrest. In many Western countries, despite adults' preferences to die at home, death will occur in a hospital. In the U.S., among those under 85 years old, 20% of all deaths occur in an intensive care unit. In Taiwan, the location of death has particular significance. Death in a hospital means that the soul will not be able to leave that setting. There is a strong preference among the Taiwanese for death to occur at home. However, among other Asian cultures, having a family member die at home brings bad luck to the household (Cheng et al., 2015).

Standard demographics such as age and health do not consistently predict patients' selected DNR options. Importantly, when patients who self-selected choices for treatment were compared with those of close relatives' choices for them, the group agreement rates were low. This pattern suggests that substituted judgment is often not a representation of the patient's preferences (Emanuel et al., 1991). One of the key difficulties in discussing treatment options in hospitalized seriously ill patients is that patients and their families often do not have an appreciation of the relative success of various treatments nor do they often understand the degree of invasiveness associated with life maintaining interventions. For example, success rates in reviving patients with cardiopulmonary resuscitation have been relatively stable—for patients younger than 70 years success rates are around 15% and approximately 12% for those older than age 70 (Schneider, Nelson & Brown, 1993). The technique has been featured on some popular medical television dramas—usually with great success. In one analysis of CPR depicted in multiple television programs, 77% of the "patients" survived the immediate arrest (Diem, Lantos, & Tulsky, 1996)—a much higher percentage than reported in most research studies. As a result of multiple influences, patients and their families, came to view CPR as a successful miracle-like intervention. CPR was soon used on a much larger scale with all patients experiencing cardiopulmonary arrest—unfortunately, the outcome of this emergency procedure was often not seen as beneficial to patients—only prolonging suffering. As CPR became more common in hospitalized intensive care unit patients, many U.S. hospitals began to develop informal processes to prevent repeated CPR when it was not seen as being in the patient's best interests—through "slow codes" or cryptic notations in patients' medical records.

In the U.S., the representation of DNR orders varies across ethnicities. Garrido, Harrington, and Prigerson (2014) found that in the U.S., non-Latino Whites were significantly more likely to have a DNR order compared with African-Americans and Latinos. In their sample of cancer patients, 44.9% of non-Latino Whites indicated a DNR order compared with 24.7% of Latinos and 19.7% of African-Americans.

Patients with DNR orders placed a higher value on comfort and pain relief than life extension (Garrido et al., 2014). African-American patients with DNR orders were significantly more likely to endorse a preference for mechanical ventilation even if it would only extend life for a day. With respect to feeding tubes, close to half of African-American patients and 40% of Latino patients indicated that they would want a feeding tube if it extended life one day compared with 26% of Whites (Garrido et al., 2014). A study of cancer patients found that African-American patients, in particular, were much more optimistic about the success of treatment—to the extent that the level of optimism about treatment success far outweighed available clinical data (Gramling et al., 2016).

Religious and spiritual orientation also plays a role in agreeing to DNR orders. Family members who are religious and believe that human life was in the hands of God would typically wish to maintain life support. "Do everything that is possible to keep him alive. And if it doesn't work, that's because God is asking for him." (Ernecoff et al., 2015, p. e-4). The religious affiliation of the physician also appears to play a role in how DNR orders are implemented. For example, withdrawal of ventilators was less common when the physician was Jewish, Muslim or of the Greek Orthodox faith (Sprung et al., 2007). Protestant physicians, by contrast, were more likely to initiate limiting treatment earlier in the hospital course than those of Greek Orthodox background. In cases where treatment was withdrawn, the time to death was longer when the physician was Jewish versus Christian Protestant (Sprung et al., 2007).

While it might appear that the PSDA would influence healthcare providers sensitivity to end-of-life issues and increase documentation of DNR orders, this pattern has not been consistently found to be the case. In a large multisite study of Medicare patients, while early DNR orders (during the first two hospital days), did increase, DNR orders occurring later in the hospital course actually declined after PSDA implementation (Baker, Einstadter, Husak, & Cebul, 2003).

5.2.1 International Perspectives on DNR Orders and Limiting Care

In the Jewish tradition, the sanctity of life is the preeminent value guiding decisions about whether to initiate and maintain artificial nutrition. Withholding nutrition, according to a minority of Jewish ethical scholars, might be considered acceptable in persons in persistent vegetative states who have a limited life expectancy (Sturman, 2003). From the perspective of Halachic law, these patients' deaths are foregone conclusions and because death is inevitable, one can justify not initiating artificial feeding. However, in Israel this view is, at present, a minority perspective. However, it is very clear that a patient would not be permitted under Halachic law to "autonomously" refuse artificial nutrition or hydration. It is not within the authority of medical professionals to intentionally cause the end-of-life either through commission (excessive sedation of a terminally ill individual) or omission such as not performing CPR. Additionally, patient autonomy is seen as secondary and actually illusory since one's life course is guided by God. Importantly, from the perspective of Judaic ethics, refusing life-sustaining treatment would be considered immoral and a violation of God's will (Sturman, 2003).

More recent interpretations of Halachic law have suggested that in situations in which reversibility of a medical condition is not possible and that the patient will likely be deceased within weeks to a year (Sturman, 2003), treatments may be withheld. Discontinuing a treatment versus not initiating an intervention are distinct actions. Importantly, within Halachic law, not initiating mechanical ventilation, CPR, or dialysis or not performing surgery may be acceptable if it is determined that these interventions will not reverse the patient's condition but only prolong the dying process. However, since feeding is a basic human function, it is not morally acceptable to withhold nutrition if one of the objectives of this deliberate omission is to expedite death. In some circumstances when it is determined that the initiation of artificial nutrition such as inserting a feeding tube would, by itself, cause potential harm to the patient, withholding this intervention is acceptable.

While it is suggested that up to 80% of North American physicians used DNR orders, these documents are less common in other countries such as Japan. In keeping with preferences in Asian cultures, a study of terminally ill patients in Taiwan found that fewer than 20% of those who signed consent documents for limiting care (DNR orders) signed it themselves (Huang, Hu, Chiu, & Chen, 2008). Nearly 80% of DNR documents were signed only by family members. In their study, Huang and colleagues (2008) noted that patients appeared quite willing to allow family members to make these decisions. Similarly, consent documents for entering hospice were, in most cases, signed only by family members and did not include the patient's signature (Huang et al., 2008). Among a sample of Chinese physicians, 53% did not agree with the practice of having DNR orders.

In Islamic law, the obligation not to initiate harm or to alleviate harm is more important than beneficence. Generally speaking, in Islamic law, intentional withholding of artificial hydration or nutrition would be considered murder (Al-Bar & Chamsi-Pasha, 2015). In Saudi Arabia, the Council of Scholars that addresses religious aspects of law does permit DNR orders if three competent physicians conclude that this is appropriate. Stopping or withdrawing treatment is permitted in cases in which a patient is clearly brain-dead, death is considered to be imminent, the treatment is considered futile and intervention will only increase the patient's suffering (Rady, Vereheidje, & Ali, 2009).

5.3   Organ Transplantation

The modern era of organ transplantation began in the 1950s with a kidney transplant between twins (Cai, 2015). Dr. Christiaan Barnard performed the first successful heart transplant in 1967. Several months earlier, the first successful transplantation of a liver from a donor occurred (Starzl et al., 1968). By the mid-1980s, organ transplants had become much more common. In the United States, because of the growing demand for organs, such as kidneys, a centralized registry was established for biologically matching organs to potential recipients. However, the waiting list for recipients has typically numbered approximately 100,000 people. It has been estimated that 18 patients on the kidney transplant waiting list die every day.

In the case of transplantation of many organs, such as hearts, the time between the donor's death and implantation of the organ in the recipient is very brief. The key issue with organ transplantation is that organs must be harvested immediately upon death. In the case of a sudden and accidental death, family members are likely to be acutely grieving the death of a loved one with responses that may include denial and anger. As such, family members may not be particularly open to consenting to organ donation. Even when the newly deceased patient's donor card provides permission to harvest the organs, clinical practice suggests that if family members are in opposition to this previous consent, organ harvesting will not take place (Cai, 2015).

There has also been concern expressed about how determinations are made between persons who are judged to be better or worse candidates for organ transplantation. From the original history of kidney dialysis in the United States when dialysis machines were very limited in number, there has been concern about the social value placed on individuals as a determinant of being an organ recipient (Searight & Meredith, 2019).

In the U.S., there have been initiatives to increase organ donation among African-Americans. Particularly with respect to kidneys, African-Americans are more likely than Whites to develop end-stage kidney disease. While nearly 36% of those on the transplant list in 1999 were African-American, only about 22% of kidney recipients were African-American. On the average, African-Americans were waiting a median of 40 months compared with 20 months for whites (Alexander & Sehgal, 1998; Epstein, Ayanian, Keogh, et al., 2000). However, in the ensuing 20 years, donor organs from deceased African-American patients have increased. However, African-Americans continue to be less likely to receive a transplanted kidney. Since it is necessary to have been on dialysis to obtain a kidney transplant, this requirement in effect reduces the African-American recipient pool since as a group, African-Americans are likely to obtain treatment for end-stage renal disease and at a later point in its course (Braithwaite et al., 2009). Additionally, African-Americans appear to be less likely to complete other required procedures for being placed on the transplant list such as undergoing a pre-transplant medical evaluation (Alexander & Seghal, 1998). While latent attitudinal discrimination is likely to play some role in this racial disparity, unequal access to specialized health care resources plays a more substantive role (Searight, 2019).

5.3.1 Organ Transplantation Internationally

Signing of donor cards or obtaining permission from relatives for harvesting a recently deceased family member's organs, as is the case in the U.S., has been called the "opt in" approach. However, countries such as Austria and Spain, have used an "opt out" approach—sometimes referred to as presumed consent. In these countries, everyone who dies is assumed to be an organ donor. While one might expect that this would increase the number of available organs significantly, there have been limitations in actual donations since physicians have been hesitant to override a family member's opposition to organ donation (Ofri, 2012). Israeli law also includes compensation for living donors including lost wages and expenses. From the perspective of the Talmud, saving a life is a priority and supersedes obedience to other commandments (Ofri, 2012). Successful passage of the legislation was likely due, in part, to public support from both Israel's imams and rabbis.

As life support technology has become more sophisticated, the criteria used to define death in a potential donor have become controversial. As noted earlier, brain death has been used as a common standard. However, the definition of brain death has varied. For example, patients who are in a persistent vegetative state and believed to be permanently unconscious may be able to breathe on their own and survive for an extended period of time with artificial feeding. The competing standard for death, accepted by conservative Jewish groups in Israel, is cessation of heart activity (Sturman, 2003). Many patients diagnosed with brain death also experience cardiac arrest. However, if life support is provided through this initial acute period, patients who are "brain dead" have lived for extended periods of time—up to at least a decade (Truog & Robinson, 2003). In Israel, organ donation is typically acceptable if the potential donor has suffered brain death, has agreed prior to death to donate their organs and the family is also in agreement. While a debatable practice, in some countries such as Iran, payment for organs does occur. In Israel, buying and selling of organs was only outlawed in 2008. While the definition of death established by the Israeli Knesset is that the person has no blood pressure, cannot breathe without external support, shows no pupillary response and is declared dead by two physicians, this definition has not been without controversy. Conservative adherents to Halachala continue to argue that cessation of the heart is the definition of death (Sturman, 2003).

In Iran, the Organ Transplantation and Brain Death Act was passed in 2000. Brain death must be verified by four physicians including a neurologist. Because of the need for organs, kidney donors may receive a financial award from the government. With the cost of the procedure paid by the government and implemented by an independent agency, the surgeons and any middlemen are kept out of the interaction. In addition, only live donors are involved (Larijani, Zhaedi, Tahani, 2004).

There has been debate among Islamic scholars about the definition of death. Those who view cadaveric donation as acceptable typically agree with the brain death definition. However, those who argue from the traditional Islamic definition of death which includes an absence of heart activity, no respiration and an absence of cerebral brain activity as well as the onset of rigor mortis, do not accept the brain death standard (Rady et al., 2009). Obviously, this rather stringent definition would prevent successful cadaveric organ harvesting. In predominantly Islamic countries, presumed consent is not accepted. Permission must be obtained from the deceased in advance or from a family representative (Rady et al., 2009).

5.4 Hospice Care

The term, "hospice" has been used in two ways. First it is a place, often made to physically appear more like a home than an institution for terminally ill patients to receive palliative care. Second, hospice is an approach to treatment in which the goal is to optimize patient comfort as well as assisting the patient to remain engaged with their family and friends. Most hospice care occurs in homes and is consistent with most U.S. patients' wishes to spend their final days at home rather than in an impersonal hospital. Research does indicate that in the U.S., White European Americans are more likely to use hospice services compared with other ethnic and cultural groups. Specifically, while ethnic minorities comprise approximately 25% of the US population, only 18% of patients receiving hospice services are non-White (Ngo-Metzger, Phillips, & McCarthy, 2008).

Entering hospice typically involves giving formal consent that active interventions to prolong life and attempts to treat the underlying terminal condition will not be implemented. In healthcare settings with formal regulations, admission to hospice is seen as appropriate for persons with less than six months to live. However, it is unusual for patients to survive any significant length of time once beginning hospice care. Approximately 20% of newly admitted hospice patients died within a week with only 6% surviving longer than two months (Ngo-Metzger et al., 2008). There are gray areas regarding this policy—for example if a patient has terminal cancer and they develop pneumonia or a urinary tract infection some physicians may view treating those illnesses as a permissible comfort measure. Others, particularly in the case of pneumonia, may view active treatment as inconsistent with hospice care. In the U.S., Medicare coverage requires that a full-time adult caregiver be present with the patient. This requirement may prevent enrolling patients in home hospice if family members are working full-time or are not close geographically.

Multiple studies conducted in the United States have found that ethnic minority groups are significantly less likely to use hospice than persons of White European background. African-Americans in the U.S. are significantly less likely to enter hospice. While comprising, 13% of the population, only 8% of hospice enrollees are African-American. Surveys of potential hospice use found that African-Americans are 20% less likely to enter hospice and once receiving hospice care, significantly more likely to prematurely discontinue hospice treatment. In the African-American community, hospice may be viewed with suspicion—a place where a loved one will not receive care and be encouraged to die.

This perception, in part, may stem from a lack of acknowledgment of the seriousness of illness. Additionally as one palliative care provider noted, "acceptance of hospice care puts people off." When viewed as a destination, hospice may carry the message that we have given up on treating you and it is time to accept and adjust to death. One provider noted that they had learned with African-American families to describe hospice as a process rather than a destination. Medications that are often used as part of palliative care and hospice such as morphine were also viewed with suspicion by some African-American patients. Morphine, used to reduce pain, can also expedite death and some African-American families, at times, expressed concern that it was a form of near euthanasia (Rhodes, Batchelor, Lee, & Halm, 2015).

In the case of Asian Americans in which respect for and care of elders is a preeminent value, hospice use is also low. In addition to the requirement that the patient give informed consent to being in hospice, family members may be distressed about the implicit message that they are "giving up" on a loved one and waiting for them to die. In many Asian cultures, this message would conflict with filial piety—the respect and care that is shown to an aging adult parent (Ngo-Metzger et al., 2008).

One exception to the reduced use of hospice by ethnic minorities is found among Filipino Americans. It has been suggested that because most Filipino Americans are somewhat westernized and are predominantly Roman Catholic, hospice is looked upon more favorably. The Catholic Church has generally been supportive of hospice and Catholic hospitals often offer hospice care. From the perspective of Catholicism, hospice care may be seen as part of the journey for an "ultimate union with God" (Ngo-Metzger et al., 2008, p. 143).

In the United States, Latinos are one of the most rapidly growing ethnic minorities. At present, Latinos comprise about 15% of the U.S. population. In contrast to White Europeans, persons of Latino(a) background were less accepting of the idea of a loved one's impending death. Even after admission to hospice, Latino caregivers were more optimistic about the outcome as illustrated by this comment: "I had a lot of faith. I thought I was lucky in that he was getting better. He was already at hospice but I did not think he was going to die. Then the day the doctor called me and told me come tomorrow because he is dying... I never thought that he was leaving me" (Kreling, Selsky, Perret-Gentil, et al., 2010, p. 5).

Kreling et al. (2010) found that most White European interviewees who had experienced hospice care for a family member found the information about the dying process to be useful. This knowledge appeared to help Whites feel more in control of the dying process. Latinos, by contrast did not appreciate the written and verbal information about the dying process and often viewed it as cruel. They were also uncomfortable with the ongoing discussions about death. Interestingly, many of the White European terminally ill patients made conscious decisions about when and how to tell family members of their diagnosis and prognosis. By contrast, Latino families were more likely to view themselves as needing to be protective of the patient. In one family, the Latina mother who was dying of cancer was told by her adult children that they were going to stop treatment so she could gain some weight and then she would receive the treatment again. One Latino family member also commented that the frequent use of the term "palliative care" was helpful in that the patient did not know what it meant and was not aware that they were in hospice. Direct Spanish translation of the term "hospice" means orphanage or "place for poor people" (Kreling et al., 2010)—not synonymous with supportive, palliative care. A large survey of recently bereaved family members—nearly all of them White Europeans—found that the quality of hospice care was rated more highly when family members experienced staff as being direct and honest with them about their loved one's prognosis (Rhodes, Mitchell, Miller, Connor, & Teno, 2008).

Hospice, as an intervention, has not developed in many parts of the world. With increased immigration from developing to developed countries, recognition of diverse perspectives towards end-of-life care will continue to be a challenge. For example, in Prato, Italy, 8–20% of the overall population is of Chinese background. However, the percentage availing themselves of home-based oncology care is less than 1% (Verna, Porzio, Galli, et al., 2016).

5.5 Physician Assisted Death

Taking autonomy to its logical conclusion are the growing number of "right to die" cases that have been heard by American, Canadian and European courts in the past 20 years. In most governmental legislation, legal rulings, and even the Catholic Church's policy, a distinction is made between intentionally acting to terminate a patient's life and the law of double effect. Many patients towards the end of life may require sedative or analgesic medication for comfort. However, this medication, depending upon the dose, can also lead to death. While some have seen this as appropriate medical care, others have seen this as a form of euthanasia. Increasing doses of pain medication for palliation with the knowledge that the medication is likely to shorten life in high doses has generally been accepted by most legal and religious authorities and is usually not considered physician-assisted suicide.

5.5.1 The United States

In the United States, physician-assisted death is regulated at the state level. Oregon was the first state in the United States to permit the practice through a citizen's referendum in 1994 which was eventually enacted in 1997. In 2008, similar laws were passed in Washington. In 2013, the Vermont legislature approved physician-assisted death with California passing a similar law in 2015. In 2009, the Montana Supreme Court upheld that a physician assisting a cognitively competent, terminally ill patient with premature death, could not be prosecuted for homicide (Ganzini, 2017). At present, nine states and the District of Columbia have legalized physician assisted suicide.

In Oregon, where physician-assisted death has been legal for over 20 years, the following regulations guide physician-assisted death: the patient must be an Oregon resident demonstrating intact decisional capacity; certified to have a life expectancy of less than six month by two physicians, drugs prescribed by a licensed physician in the state, and the patient must be able to self-administer the medication. Between 1998 and 2015, a total of 1,545 prescriptions were written and 64% of patients ended their lives with the medication. The pattern of approximately one third of patients not taking the lethal medication has been fairly consistent over time. There was originally concern that physician-assisted suicide would occur with a very high rate, it actually accounts for a miniscule number of annual deaths—38.6 per 10,000 total deaths in Oregon (Blanke, LeBlanc, Hershman, Ellis, & Meyskens, 2017). This figure is much smaller than for either the Netherlands or Belgium—two other regions with longer standing laws permitting physician assisted death. At the time of their death, over 90% of the Oregon patients were enrolled in hospice program the most common diagnosis for the Oregon patients, like those in Canada and the Netherlands, was cancer.

5.5.2 Canada

In 2015, the Canadian Supreme Court in evaluating whether patients had a right to a dignified death, examined the conflict between provincial criminal law which prevents assisting any type of suicide even at the patient's request and the federal Canadian Charter of Rights and Freedom which guarantees "...the right to life liberty and security of the person..." (Attaran, 2015). The court argued that the criminal code could be overridden since it was an obstacle to individual decisions about someone's own death and bodily integrity—thus limiting the individual's right to liberty as specified in the Charter. The 2015 law does allow patients to request palliative sedation, and to refuse artificial nutrition and hydration as well as request removal of life-sustaining medical treatment or equipment such as a respirator. In order to receive physician-assisted dying, the patient must have a "grievous" and untreatable medical condition with a "reasonably foreseeable death" (Attaran, 2015). Patients also have the choice between palliative care and physician-assisted death.

Receiving assistance in dying was officially passed by the Canadian Parliament in June, 2016. There still remained a lack of clarity regarding implementation. Medical Assistance in Dying (MAiD), unlike home-based programs, occurs within a hospital. Patients who receive the service through one of Canada's four teaching hospitals are given lethal medication intravenously only after physicians conduct independent evaluations which assess "the patient's prognosis, suffering, and capacity to provide informed consent" (Li et al., 2017). Upon MAiD's implementation, the majority of patients who initially requested assistance with dying were diagnosed with cancer. However, among patients with other conditions was a patient with major depressive disorder. Initially, psychiatric conditions were "potentially eligible" for the service but in June 2016, mental health problems, alone, were no longer recognized as meeting MAiD standards (Li et al., 2017).

In the initial months after passage, 86% of Canadian patients' requests for assistance in dying were approved. Of those patients approved, three-quarters have received the assisted death protocol. As noted, in the original MAiD legislation, it was not required that the patient have a terminal physical condition. For a period of some months until the legislation was revised, patients requesting assistance in dying could do so on the basis of intractable psychiatric illness (Attaran, 2015). With the acceptance of physician-assisted suicide for mental health conditions in the Netherlands, it is highly likely that the Canadian government will be revisiting this issue. A consistent finding in the United States and the Netherlands as well as more recently in Canada is that patients requesting assistance in dying do so primarily because of psychological distress regarding current or anticipated loss of autonomy and a desire to experience control over their death (Li et al., 2017). Attaran (2015) suggests that the Canadian right to die legislation will be used as a model for countries such as South Africa, India and England which are currently examining the practice.

5.5.3 Europe

While the European Union has greatly reduced border controls between European countries, policies regarding a patient's right to die are variable within the European Economic Community. The Netherlands and Belgium have among the most liberal policies around physician-assisted death in the world. Along with Oregon in the United States, these countries have been considered "bellwethers" of the movement for assisted dying.

Great Britain: Great Britain has been particularly steadfast in not allowing the right to die for patients who have terminal illnesses or who are in minimally conscious states for extended periods of time. Britain has established a "Court of Protection" which addresses appeals for PAS and related issues (Guardian, 2011). Historically, the Court of Protection has held its hearings in secret and according to some analysts, has the power to impose experimental treatment on non-consenting patients. Previously, the Court had dealt almost exclusively with mental capacity questions particularly regarding finances or patients who refused medical treatment.

In October 2017, Noel Conway, who had a motor neuron disease, had his request for assisted dying refused by the High Court. The Court indicated that Mr. Conway could bring about his own death himself. Mr. Conway's argument was that since he estimated he had about six months to live he wished for a dignified death. Instead, with the Court ruling, his option was to remove his ventilator and with the aid of sedative medication, suffocate to death. Parliament in 2015 had rejected legislation in England and Wales to allow any type of assisted dying (BBC, 2017).

France: In France, passive euthanasia (e.g., refusing a respirator or feeding tube) has been legal since 2005. However, for both French physicians and patients, there is some ambiguity about the legality of the principle of double effect. The 2005 Lunette Law does allow physicians to provide a range of measures for symptom control even when they may shorten the patient's lifespan. However, in a 2007 case involving the death of a 65-year-old woman with terminal pancreatic cancer, a nurse and physician were charged with deliberately poisoning the patient (Sokol, 2007). The French criminal court acquitted the nurse who had administered potassium chloride, while the physician who wrote the prescription received a one-year suspended sentence. Both health care professionals were charged with deliberately poisoning the patient. In 2016, additional legislation was passed which more explicitly permitted cessation of artificial hydration and nutrition as well as the patient's right to have continuous deep sedation until death. Deep sedation is permitted when the patient has a serious and incurable condition with a short life expectancy or the patient's decision to stop the treatment would lead to significant suffering in the context of brief life expectancy (Raus, Chambaere, & Sterckx, 2016).

5.5.4 The Netherlands and Belgium

The Netherlands and Belgium have permitted physician suicide for over a decade. Euthanasia, as it is termed, became legal in the Netherlands and Belgium in 2002. These countries are considered to have the most liberal laws on physician assisted death in the world. The use of PAS has become increasingly common although certainly not the most common cause of death. In the Netherlands, the decision not to prosecute physicians who assisted patients in ending their lives dates back to 2002. However, in 2012, after physician-assisted death had not been prosecuted for a decade, one in 30 people in the Netherlands died by euthanasia. This represents a threefold increase since the practice was introduced. In Belgium, the 2013 PAS rate was 4.6%—one in 22 deaths. Available data also suggest that Belgian physicians are increasingly authorizing PAS—authorization rates in 2007 were 55% but grew to 77% in 2013 (Lerner & Caplan, 2015).

Belgium and the Netherlands also have come to represent the fears about PAS held by physicians, ethicists, as well as some segments of the general population. In a study of an end-of-life clinic in the Netherlands, nearly 7% of those who were granted the availability of PAS reported that a major motivator for their request was that they were "tired of living." While not the sole reason for requesting PAS, 49% of the requests cited loneliness is a factor (Lerner & Caplan, 2015).

In Belgium, cases involving euthanasia and or physician-assisted dying have included conditions such as autism, anorexia nervosa and chronic fatigue syndrome. Lerner and Caplan (2015) raise concern that patient self-determination may be such a high priority in the healthcare community that "The risk now is that people no longer search for a way to endure their suffering" (Ross, 2015 cited in Lerner & Caplan, 2015, p. e-2). In other words, are countries such as the Netherlands and Belgium turning to physicians to solve with euthanasia what are essentially psychosocial issues? (Lerner & Caplan, 2015, p. E2).

As noted earlier in the discussion of historical change in family dynamics, Belgium now permits physician-assisted death for minors. Consent of both parents and the minor patient is necessary for assisted death to occur. Commentators in the United States have expressed concern that Belgium and the Netherlands are "pushing" the boundaries of physician-assisted death such that it is being presented as an option for non-terminal health conditions such as degenerative diseases (Cohen-Almagor, 2017).

PAS and Psychiatric Illness: A practice that has become quite controversial is the implementation of PAS for psychiatric disorders. In the few reported cases where this has occurred, the patient did not have significant medical illness, death due to physical causes was unlikely to occur in the foreseeable future, and the patient was determined to be competent to make the request. In the U.S., While PAS is legal in some states, there has been generally strong opposition to extending the option to patients whose only illness is psychiatric. Opponents of the practice indicate concern that the hopelessness associated with the patient's request may reflect transient distorted thinking. Miller and Appelbaum (2018) also point out that PAS for psychiatric conditions could potentially be requested more in the United States than in countries with universal health coverage because of our fragmented and often inaccessible mental health system. Because of inaccessibility to optimal treatment for conditions such as major depressive disorder, patients may request premature death as an option. Finally, in the United States, it is illegal for individuals to kill themselves and the police are required to intervene. Allowing PAS for psychiatric conditions would appear to conflict with the criminalization of suicide.

Miller and Appelbaum (2018) describe the case of a 64-year-old Belgian woman who had a long history of depression and who had requested PAS after the breakup of a relationship. She apparently sought PAS from physicians but could not find two physicians who agreed that she had an incurable illness. While the patient's social situation was worsened by the estrangement from her two children, a therapist offered the opinion that these rejections would prevent her psychiatric condition from improving. Of note, the patient's treatment history was limited to psychotherapy and antidepressant medication. Interventions such as electroconvulsive therapy, transmagnetic stimulation and implantable deep brain stimulation devices had not been attempted. Additionally, the patient's adult children apparently were not contacted or involved in the mother's decision.

In response to Miller and Appelbaum (2018), Vandenberghe (2018), a Belgian psychiatrist, argued that while PAS for psychiatric conditions should not be a common practice, there are patients who continue to suffer and do not respond to a range of evidence-based treatments. In Belgium and the Netherlands, rather than requiring that patients have a terminal condition, physicians in these countries use the standard that "patient should be able to end irremediable and unbearable suffering caused by an illness for which treatment has been futile" (Vandenberghe, 2018, p. 885). Vandenberghe notes that these patients often report a very poor quality of life and given the failures of psychiatric treatment, see no end to this emotionally painful existence. Vandenberghe (2018), however, recommends a much more rigorous process for euthanasia for psychiatric conditions than for terminal medical conditions. He recommends that a committee of mental health professionals conduct a multidisciplinary evaluation before granting the patient's request. The committee would review the patient's history and treatment, include attention to the patient's life context, and interview the patient's family.

5.6 Conclusion

As the human lifespan lengthens and medical technology continues to develop, we are faced with an array of choices to prolong or end our lives. We can communicate our preferences with legally prepared documents providing instructions about whether we would like various interventions to keep us alive when we are seriously ill. It is possible to, at least in theory, prevent physicians from using heroic measures to keep us alive. Additionally, hospice and palliative care are designed to help us have a "good", minimally painful, death. We can even perform acts of altruism in our last days by indicating in advance that we would like to donate our organs upon our death. However, while on the surface, these choices may appear desirable, electing any of these options often reflects important underlying values regarding life, commitments to others, and views of suffering.

References

Alexander, G. C., & Sehgal, A. R. (1998). Barriers to cadaveric renal transplantation among blacks, women, and the poor. *JAMA, 280*(13), 1148–1152.

Al-Bar, M. A., & Chamsi-Pasha, H. (2015). *Contemporary bioethics: Islamic Perspective*. New York: Springer.

Attaran, A. (2015). Unanimity on death with dignity—Legalizing physician-assisted dying in Canada. *New England Journal of Medicine, 372*(22), 2080–2082.

Baker, D. W., Einstadter, D., Husak, S., & Cebul, R. D. (2003). Changes in the use of do-not-resuscitate orders after implementation of the Patient Self-Determination Act. *Journal of General Internal Medicine, 18*(5), 343–349.

BBC (2017) Terminally ill Noel Conway Loses Supreme Court Appeal. https://www.bbc.com/news/uk-england-shropshire-46359845.

Beširević, V. (2010). End-of-life care in the 21st century: Advance directives in universal rights discourse. *Bioethics, 24*(3), 105–112.

Blackhall, L. J., Murphy, S. T., Frank, G., Michel, V., & Azen, S. (1995). Ethnicity and attitudes toward patient autonomy. *JAMA, 274*(10), 820–825.

Blanke, C., LeBlanc, M., Hershman, D., Ellis, L., & Meyskens, F. (2017). Characterizing 18 years of the death with dignity act in Oregon. *JAMA Oncology, 3*(10), 1403–1406.

Braithwaite, R. L., Taylor, S. E., & Treadwell, H. M. (2009). *Health issues in the Black community*. New York: John Wiley & Sons.

Cai, Y. (2015). On family informed consent: On the legislation of organ donation in China. In R. Fan (Ed.), *Family-oriented informed consent: East Asian and American perspectives* (pp. 187–200). New York: Springer.

Chan, H. M., Doris, M. T., Wong, K. H., Lai, J. C. L., & Chui, C. K. (2015). End-of-life decision making in Hong Kong: the appeal of the shared decision making model. In *Family-Oriented Informed Consent* (pp. 149–167). Cham: Springer.

Cheng, S. Y., Suh, S. Y., Morita, T., Oyama, Y., Chiu, T. Y., Koh, S. J., ... Tsuneto, S. (2015). A cross-cultural study on behaviors when death is approaching in East Asian countries: What are the physician-perceived common beliefs and practices? *Medicine, 94*(39), 1–5.

Cohen-Almagor, R. (2017). Euthanizing people who are 'Tired of Life' in Belgium. *Euthanasia and Assisted Suicide: Lessons from Belgium* (pp. 188–201). Cambridge: Cambridge University Press.

Diem, S. J., Lantos, J. D., & Tulsky, J. A. (1996). Cardiopulmonary resuscitation on television—Miracles and misinformation. *New England Journal of Medicine, 334*(24), 1578–1582.

Emanuel, L. L., Barry, M. J., Stoeckle, J. D., Ettelson, L. M., & Emanuel, E. J. (1991). Advance directives for medical care—A case for greater use. *New England Journal of Medicine, 324*(13), 889–895.

Ernecoff, N. C., Curlin, F. A., Buddadhumaruk, P., & White, D. B. (2015). Health care professionals' responses to religious or spiritual statements by surrogate decision makers during goals-of-care discussions. *JAMA Internal Medicine, 175*(10), 1662–1669.

Epstein, A. M., Ayanian, J. Z., Keogh, J. H., Noonan, S. J., Armistead, N., Cleary, P. D., ... Conti, R. M. (2000). Racial disparities in access to renal transplantation—clinically appropriate or due to underuse or overuse?. *The New England Journal of Medicine, 343*(21), 1537.

Frankford, D. M. (2015). The remarkable staying power of "death panels". *Journal of Health Politics, Policy and Law, 40*(5), 1087–1101.

Ganzini, L. (2017). Legalized physician assisted death in Oregon—Eighteen years' experience. In *Assistierter Suizid: Der Stand der Wissenschaft* (pp. 7–20). Berlin, Heidelberg: Springer.

Garrido, M. M., Harrington, S. T., & Prigerson, H. G. (2014). End-of-life treatment preferences: A key to reducing ethnic/racial disparities in advance care planning?. *Cancer, 120*(24), 3981–3986.

Gramling, R., Fiscella, K., Xing, G., Hoerger, M., Duberstein, P., Plumb, S., ... Epstein, R. M. (2016). Determinants of patient-oncologist prognostic discordance in advanced cancer. *JAMA Oncology, 2*(11), 1421–1426.

Huang, C. H., Hu, W. Y., Chiu, T. Y., & Chen, C. Y. (2008). The practicalities of terminally ill patients signing their own DNR orders—a study in Taiwan. *Journal of Medical Ethics, 34*(5), 336–340.

Johnson, K. S., Kuchibhatla, M., & Tulsky, J. A. (2008). What explains racial differences in the use of advance directives and attitudes toward hospice care?. *Journal of the American Geriatrics Society, 56*(10), 1953–1958.

Kreling, B., Selsky, C., Perret-Gentil, M., Huerta, E. E., Mandelblatt, J. S., & Latin American Cancer Research Coalition. (2010). 'The worst thing about hospice is that they talk about death': Contrasting hospice decisions and experience among immigrant Central and South American Latinos with US-born White, non-Latino cancer caregivers. *Palliative Medicine, 24*(4), 427–434.

Larijani, B., Zahedi, F., & Taheri, E. (2004). Ethical and legal aspects of organ transplantation in Iran. In *Transplantation proceedings* (Vol. 36, No. 5, pp. 1241–1244). New York: Elsevier.

Lee, I. (2015). Filial duty as the moral foundation of caring for the elderly: its possibility and limitations. In R. Fan (Ed.), *Family-oriented informed consent: East Asian and American perspectives* (pp. 137–147). New York: Springer.

Lerner, B. H., & Caplan, A. L. (2015). Euthanasia in Belgium and the Netherlands: On a slippery slope? *JAMA Internal Medicine, 175*(10), 1640–1641.

Li, M., Watt, S., Escaf, M., Gardam, M., Heesters, A., O'Leary, G., & Rodin, G. (2017). Medical assistance in dying—Implementing a hospital-based program in Canada. *New England Journal of Medicine, 376*(21), 2082–2088.

Miller, F. G., & Appelbaum, P. S. (2018). Physician-assisted death for psychiatric patients—Misguided public policy. *New England Journal of Medicine, 378*(10), 883–885.

Morrell, E. D., Brown, B. P., Qi, R., Drabiak, K., & Helft, P. R. (2008). The do-not-resuscitate order: associations with advance directives, physician specialty and documentation of discussion 15 years after the Patient Self-Determination Act. *Journal of Medical Ethics, 34*(9), 642–647.

Ngo-Metzger, Q., Phillips, R. S., & McCarthy, E. P. (2008). Ethnic disparities in hospice use among Asian-American and Pacific Islander patients dying with cancer. *Journal of the American Geriatrics Society, 56*(1), 139–144.

Nicholas, L. H., Langa, K. M., Iwashyna, T. J., & Weir, D. R. (2011). Regional variation in the association between advance directives and end-of-life Medicare expenditures. *Jonal of the American Medical Association, 306*(13), 1447–1453.

Ofri, D. (2012). In Israel, a new approach to organ donation. *The New York Times*. February 16.

Peinado-Gorlat, P., Castro-Martínez, F. J., Arriba-Marcos, B., Melguizo-Jiménez, M., & Barrio-Cantalejo, I. (2015). Roma women's perspectives on end-of-life decisions. *Journal of Bioethical Inquiry, 12*(4), 687–698.

Perkins, H. S. (2007). Controlling death: the false promise of advance directives. *Annals of Internal Medicine, 147*(1), 51–57.

Rady, M. Y., Verheijde, J. L., & Ali, M. S. (2009). Islam and end-of-life practices in organ donation for transplantation: New questions and serious sociocultural consequences. *HEC Forum, 21*(2), 175.

Raus, K., Chambaere, K., & Sterckx, S. (2016). Controversies surrounding continuous deep sedation at the end of life: The parliamentary and societal debates in France. *BMC Medical Ethics, 17*(1), 36.

Rhodes, R. L., Batchelor, K., Lee, S. C., & Halm, E. A. (2015). Barriers to end-of-life care for African Americans from the providers' perspective: Opportunity for intervention development. *American Journal of Hospice and Palliative Medicine, 32*(2), 137–143.

Rhodes, R. L., Mitchell, S. L., Miller, S. C., Connor, S. R., Teno, J. M. (2008). Bereaved family members' evaluation of hospice care: What factors influence overall satisfaction with services? *Journal of Pain and Symptom Management, 35*(4), 365–371.

Ross W. (2015). Dying Dutch: Euthanasia spreads across Europe. *Newsweek*. February 12.

Schneider, A. P., Nelson, D. J., & Brown, D. D. (1993). In-hospital cardiopulmonary resuscitation: A 30-year review. *The Journal of the American Board of Family Practice, 6*(2), 91–101.

Searight, H. R. (2019). *Conducting psychotherapy with an interpreter*. Oxford Handbook of Psychotherapy Ethics, in press.

Searight, H. R., & Gafford, J. (2005). "It's like playing with your destiny": Bosnian immigrants' views of advance directives and end-of-life decision-making. *Journal of Immigrant Health, 7*(3), 195–203.

Searight, H. R., & Meredith, T. (2019). Physician deception and telling the truth about medical "Bad News": History, ethical perspectives, and cultural issues. In *The Palgrave Handbook of Deceptive Communication* (pp. 647–672). New York: Palgrave Macmillan.

Silveira, M. J., Kim, S. Y., & Langa, K. M. (2010). Advance directives and outcomes of surrogate decision making before death. *New England Journal of Medicine, 362*(13), 1211–1218.

Sokol, R. (2007, March 21). Essay: The right to die. *New York Times*.

Sprung, C. L., Maia, P., Bulow, H. H., Ricou, B., Armaganidis, A., Baras, M., ... & Nakos, G. (2007). The importance of religious affiliation and culture on end-of-life decisions in European intensive care units. *Intensive Care Medicine, 33*(10), 1732–1739.

Starzl, T. E., Groth, C. G., Brettschneider, L., Penn, I., Fulginiti, V. A., Moon, J. B., ... Porter, K. A. (1968). Orthotopic homotransplantation of the human liver. *Annals of Surgery, 168*(3), 392.

Sturman, R. L. (2003). *Six lives in Jerusalem: End-of-life decisions in Jerusalem: Cultural, medical, ethical and legal considerations*. New York: Springer.

The Guardian (2011, September, 28). Right to die case: How Britain's most secretive court operates.

Truog, R. D., & Robinson, W. M. (2003). Role of brain death and the dead-donor rule in the ethics of organ transplantation. *Critical Care Medicine, 31*(9), 2391–2396.

Vandenberghe, J. (2018). Physician-assisted suicide and psychiatric illness. *New England Journal of Medicine, 378*(10), 885–887.

Verna, L., Porzio, G., Galli, B., Sacco, I., Brogi, L., Spinelli, G., & Giusti, R. (2016). Immigrants accessing end-of-life care in Italy: The tuscany tumor association experience. *Journal of Pain and Symptom Management, 51*(5), e7.

Yadav, K. N., Gabler, N. B., Cooney, E., Kent, S., Kim, J., Herbst, N., ... Courtright, K. R. (2017). Approximately one in three US adults completes any type of advance directive for end-of-life care. *Health Affairs, 36*(7), 1244–1251.

Yang, Y. (2015). A family oriented Confucian approach to advance directives in end-of-life decision-making for incompetent elderly patients. In R. Fan (Ed.), *Family-oriented informed consent: East Asian and American perspectives* (pp. 257–270). New York: Springer.

Chapter 6
Why Is There Such Diversity in Preferences for End-of-Life Care? Explanations and Narratives

The differences in views of advance directives, hospice care and do not resuscitate orders that characterize different ethnic and cultural groups may appear illogical to healthcare professionals of White European background. However, these patterns reflect core cultural values around the family, one's duty to others, the meaning attributed to suffering, views of language as having the power to shape events, and multiple centuries of history with a healthcare system perceived as exploitative and abusive. This chapter attempts to provide some explanatory background about the meaning of these differences among cultures and ethnicities toward end-of-life care.

6.1 Saying It Makes Death Real

The example at the beginning of the first chapter and the description of interpreter-mediated communication highlights several issues that occur with Native American populations. A 70-year-old man of Canadian aboriginal background is being evaluated for possible prostate cancer. Prior to undergoing testing, a young female interpreter struggles to fulfill her formal role as an interpreter while maintaining cultural beliefs. In addition to experiencing emotional difficulty discussing male genitalia with an elder in the community, the interpreter refuses to convey the concept of cancer but, instead, uses the term "growth." In a culture in which elders are treated with respect, a young woman is not the optimal choice of an interpreter for this situation. Additionally, she does not interpret the term "cancer" out of the belief that "speaking the future may bring it to pass." (Ellerby, McKenzie, McKay, Gariépy, & Kaufert, 2000, p. 849).

After the tests have been completed, the patient, his adult son and a male interpreter have a follow-up appointment with the physician. During the visit, an oncologist explains that the patient has advanced cancer that is spreading to the bone. Because of the late stage of the cancer, the physician recommends palliative rather

© The Author(s), under exclusive license to Springer Nature Switzerland AG 2019
H. R. Searight, *Ethical Challenges in Multi-Cultural Patient Care*,
SpringerBriefs in Ethics, https://doi.org/10.1007/978-3-030-23544-4_6

than curative treatment. As the interpreter begins to convey the physician's diagnostic and prognostic information, the son interrupts and corrects the interpreter for using the Ojibwa term for cancer which means "being eaten from within." The son goes on to tell the interpreter that giving his father this information will only make his father distressed and "bring death closer" (Ellerby et al., 2000). After a family consultation including a sharing circle, a group healing method including attention to factual content as well as spiritual and emotional dimensions (Lavallée, 2009), with the caregiver's family and the patient, the family indicates that they will gradually provide information to their husband and father over time.

As Caresse and Rhodes' (1995) description of introducing the PSDA on the Navajo reservation makes clear, both advance directives and living wills ask us to imagine what our values would be if we were in a state in which our chances of survival were poor or, at best, unknown. In cultures in which language, thought and action are inextricably linked, end of life discussions are not hypothetical future possibilities but, instead, give terminal illness and death a reality and possibly make it a self-fulfilling prophecy. In the Caresse and Rhodes (1995) study, a number of their Native American informants would not even discuss advance care planning because of the inherent danger in speaking aloud about death and disability. This perspective, while not as explicit as among the Navajo, has been found in other cultures such as recent Bosnian immigrants to the U.S. who reported a similar reluctance to discuss death (Searight & Gafford, 2005). Rhodes, Batchelor, Lee, and Halm (2015) noted that among African-American patients, recognizing that they had an illness was seen as accelerating their death. This issue has also been mentioned in discussions of advance directives among African-Americans. A palliative care social worker mentioned: "I see that there is sometimes a superstition about writing an advance directive because I've heard "if I write it down, it makes it happen." (Rhodes et al., 2015, p. 140).

While there is certainly evidence that minorities have more difficulty accessing the healthcare system and may receive less aggressive care, the difficulty acknowledging serious illness may be another factor contributing to the delay in seeking treatment until the condition is fairly far advanced. A palliative care physician observes: "Sometimes you have people who will not even admit they're sick. I mean, I've had patients with fumigating breast wounds that come in with duct tape and paper towels around their chests that say, 'I can't own this.' This is the hardest challenge for me—the hardest thing for me to take care of" (Rhodes et al., 2015, p. 140).

In Korea and Taiwan, the power of language also appears to be a significant factor making even family members reluctant to discuss death among themselves. Similar to the pattern described for some Native American communities in the United States, there is a view in some Asian cultures, that saying something "out loud" makes it more likely to occur (Cheng et al., 2015). The concept of acknowledging the family member's likely future death is seen in and of itself as making that death more malevolent as in this saying: "a bad life is better than a good death" (Cheng et al., 2015, p. 5).

Physicians' discussions with surrogate decision-makers and patients about the likely outcome of their illness can be framed linguistically in optimistic or pessimistic terms. As one surrogate put it, if the doctor says there's a 5% chance of survival you say to yourself that they have a 5% chance and forget the 95% (Zier et al., 2008, p. 6). Given that the only other option to the 5% chance is death, the figure, while small, is imbued with optimism. Research in other areas has shown that patients interpret percentages differently than they do odds. For example, saying that a patient has a 5% chance of survival is perceived as a worse outcome than saying that they have a one in 20 chance of survival.

6.2 The African-American Community and the Health Care System: The Origins of Mistrust

As noted earlier, in the United States, African-Americans are less likely to develop advance directives, use hospice services, serve as organ donors and authorize a do-not-resuscitate order. There are multiple reasons that have been offered for this persistent difference. As will be discussed below, African-American patients, as a group, do not receive the same level of medical care as Whites. While the officially segregated hospitals and healthcare clinics that were present in the southern United States up until the 1960s no longer exist, there is evidence that the amount and of medical care received by African-Americans is still not equivalent to that of Whites (Searight, 2019). Additionally, there is a history of experimentation conducted with African-Americans that was deceptive and for which informed consent was not obtained. Much of this history is striking in its inhumanity. While African-American patients may not be able to always state the specific historical events involved, the narrative of harm and being taken advantage of by the White medical establishment is well known and undoubtedly influences interactions between African American patients and health care professionals.

One of the targets of public health efforts in the United States is a reduction in health disparities. It is very clear that certain groups within the U.S. population have a higher burden of disability and death than others. Similarly, disparities extend to healthcare, itself. Certain populations within the United States—often people of color—are less likely to receive more sophisticated surgical procedures, vaccinations, adequate pharmacotherapy, and appropriate prenatal care.

Disparities in health status among African-Americans versus Whites when placed in the context of history are sadly understandable. African-Americans came to the United States primarily as slaves with death rates during transport estimated to be as high as 35% (Noonan, Velasco-Mondragon, & Wagner, 2016). After 250 years of slavery with poor living conditions, little medical care, and significant poverty, emancipation occurred. However, African-Americans have continued to live with a history of discrimination.

In the United States, there is considerable evidence that African-Americans as a group, have poorer health status and reduced access to quality healthcare. African-Americans die at younger ages. In 2014, the U.S. life expectancy overall was 76.4 years for males and 81.2 years for females. However, for African-Americans at birth, life expectancy was 72 years for males and 78.1 years for females. This difference in mortality rates begins at birth–infant mortality rates have consistently been two and half times greater for African-Americans compared with Whites. In the past two decades, African-American women have been about 10 to 20% less likely than whites to obtain prenatal care during the first trimester of their pregnancy (Noonan et al., 2016).

The single greatest factor in contributing to differences in White European and African-American mortality is the prevalence of cardiovascular disease. A now well-known study in the late 1990s demonstrated that unrecognized racial bias likely influences medical care (Schulman et al., 1999). In the study, a sample of over 700 physicians viewed a video interview and were given results of a thallium stress test and EEG test results as well as background history on the symptoms of a hypothetical patient. Based on the information provided, the physician was to make recommendations. There were four vignettes: a White female, an African-American female, a White male and an African-American male—all between the ages of 55 and 70 years. The African-American patients—particularly women—were less likely to be referred for cardiac catheterization. This difference remained even after controlling for symptoms. It was suggested that this pattern of findings was not due to overt racial bias but more likely to be a function of "subconscious perceptions." which activated a "cultural stereotype." (Schulman et al., 1999, pp. 624–625). Research reviews indicate that African-Americans are still less likely to receive cardiac catheterization, medications to decrease blood pressure and stroke risk, as well as less likely to receive coronary artery bypass graft surgery (Kressin & Peterson, 2001).

Cancer treatment also appears to be significantly less aggressive when patients are African-American. Compared with White women. African-American women with breast cancer were more likely to undergo the greater disfigurement associated with mastectomy and less likely to receive less intrusive surgical procedures and radiation therapy. African-Americans were also significantly less likely to undergo surgery for non-small cell lung cancer. Surgical intervention is known to decrease mortality rates for the condition (Washington, 2006).

Finally, in a review of quality of care received by race, African-Americans were significantly more likely to receive fewer desirable interventions—limb amputation rather than efforts to conserve the arm or leg (Lavery, Ashry, Van Houtum, et al., 1996). Additionally, compared with Whites, African-Americans as well as Hispanics were likely to receive a poorer quality of care across a range of conditions from cancer to mental health (Schneider, Zaslavsky, & Epstein, 2002).

6.2.1   A History of Medical Exploitation

Up until the early 1900s, medical schools were generally proprietary institutions of varying quality. However, a factor that became important was the availability of cadavers for anatomical dissection and instruction. These cadavers were difficult to acquire. The South Carolina Medical College, in advertising their institution, emphasized the large number of cadavers available: "subjects being obtained from the colored population in sufficient numbers by every purpose, and proper dissection carried out without offending any individuals in the community" (Washington, 2006; cited in Halperin, 2007). In the 1860s, disturbing or in any way mutilating a cadaver was seen as a form of desecration that would make it difficult for one's' soul to survive in the afterlife (Halperin, 2007). As a result, a trade developed in robbing graves. Those who had the resources would often build a protective wrought iron fence around a relative's grave to protect their relative's body from intruders.

Additionally, when a slave died, their body was often sold to a medical school. A notice from a Dr. T Stillman even sought the bodies of slaves before they were deceased. Specifically, he requested slaves with incurable conditions and listed a number of illnesses including diseases of the liver, bladder and even those with diarrhea and dysentery (Washington, 2006).

In the African–American community, narratives of night doctors and resurrectionists digging up graves, acquiring cadavers, and selling them to medical schools, became well-known (Halperin, 2007). There is evidence that Black bodies were more likely to be used in medical schools almost exclusively serving White students. One medical institution that relied heavily upon cadavers was the Medical College of Georgia. In 1852, the College purchased a slave and assigned him the task of obtaining cadavers. He apparently drew heavily upon the local Cedar Grove Cemetery with a predominantly indigent Black population (Halperin, 2007). In 1989, during renovation of the medical college a number of human bones were found. During systematic excavation guided by an anthropologist, a total of 9000 bone fragments were obtained. It was estimated that about 80% of the skeletal remains were from African-Americans. During the 1930s when there was large migration of southern Blacks to the large cities of the northern United States, southern Whites, concerned about losing a population about whom upon whom they were reliant for much manual labor, reportedly started rumors of night doctors (Washington, 2006). Night doctors reportedly roamed African-American neighborhoods in large northern cities and looked for victims to become salable cadavers.

J. Marion Sims, often described as one of the greatest surgeons of the 1800s as well as the father of modern gynecology, developed many of his surgical procedures with slave women—some of whom he specifically purchased for this purpose. One of his early surgical contributions was repair of fistulas which led to painful birth and both urinary and fecal incontinence. He also performed these procedures on local slave women for their owners (Petros, Abdenstein, & Browning, 2018). Anesthesia at the time was not well developed and Sims reportedly performed these procedures without any form of anesthetic (Washington, 2006). There are some reports that Sims deliberately induced fistulas in at least one of the slave women to develop his surgical repair technique (Washington, 2006). Sims became quite successful and well-known. He then moved to New York City and is buried there. A statue of J. Marion Sims, until very recently, was in Central Park. Because of protests, it was removed from that site.

The Tuskegee syphilis study that lasted over 40 years is believed to play a major role in African-American patients' reluctance to cease active treatment and move to palliative care for end-of-life conditions. In the study, conducted by the U.S. Public Health Service, African-American men in the rural southern United States who had syphilis were followed to study the course of the disease. The men, many of whom were illiterate, were promised free meals and free burials. They also believed that the medical tests that they were undergoing as part of the study, including lumbar punctures, were treatments for their condition. The study began in 1930; 10 years after its initiation it was established that penicillin was effective in treating syphilis. The men were never informed of this treatment option and the U.S. Public Health Service investigators continued to follow the men and document the course of the illness (Reverby, 2009).

The syphilis study continued until 1972 when a social worker for the Public Health Service expressed concern about the study to his superiors (Reverby, 2009). When they did not respond, he informed the New York Times about the study. At that time, the study was abruptly stopped. However, despite congressional hearings, the surviving men did not receive a formal apology from the U.S. government until President Clinton offered one in the early 1990s.

Since that time there have been several studies in which African-Americans were disproportionately represented in medical investigations in which they did not give complete informed consent and which posed some risks. For example, in the late 1980s a new measles vaccine was administered to African-American and Hispanic children in the Los Angeles area. Parental consent was not given and in particular, parents did not know that the vaccine being tested was sometimes given in doses 500 times the appropriate level (Washington, 2006).

Washington (2006) describes the practice of conducting uninformed sterilizations on African-American women-primarily in the southern United States. At times, the hysterectomy was an "add on" to another medical procedure. In some instances, it was presented as a form of contraception while other, less invasive methods were downplayed or not discussed (Washington, 2006).

6.2.2   Mistrust as a Legacy

This extended history of slavery, discrimination and deception in medical research, has led many African-Americans to be mistrustful of the healthcare system. When palliative care is raised as an option with terminal illness, a not uncommon reaction is that the patient is being given up on and not provided the same level of care that a middle class White patient would receive. One palliative care physician noted that sometimes African-American patients will "come in swinging" but acknowledged that he could understand this stance because of the history of maltreatment of African-American patients by the U.S. medical system. Given this history, it is understandable that many African-American patients when they deal with the healthcare system may come across as adversarial. There is concern that family members are not going to receive adequate care and the need to "come in swinging"... stems from "... doing what they have to do to look out for their family" (Rhodes et al., 2015, p. 6).

Today, African-Americans report lower levels of satisfaction with their medical care (Searight, 2019). Family members also report lower levels of satisfaction in their interaction with physicians-particularly around end-of-life care. A recent study attempted to quantify mistrust. Not surprisingly, African-Americans scored higher on the index of mistrust of the medical system than Whites (Boag, Suresh, Celi, Szolovits, & Ghassemi, 2018). Additionally, the mistrust index was positively associated with more aggressive end-of-life care—suggesting that patients and their families who are less trusting of their health care providers are more likely to reject palliative care and instead request and receive aggressive treatment, considered futile by some physicians, until death (Boag et al., 2018).

African-American patients may be more likely to view receipt of even routine medical care as an experiment (Jacobs et al., 2006). Narratives from a recent focus group study on African-American patients' use of healthcare included explicit references to Tuskegee: "It reminds me of the Tuskegee Institute where they messed around and they made the brothers have the disease instead of treating them... they just wanted to see how it was going to affect them. So maybe sometimes you go, instead of getting treated, they just want to see what it's going to do to you and I'll try this and try that and they may even give you a sugar pill" (Jacobs, Rolle, Ferrans, Whitaker, & Warnecke, 2006, p. 645).

In another study, while not mentioning Tuskegee by name, respondents expressed concern that African-American patients, at the end of their lives, would be unknowingly used for research: "That is just like experiments. People experimenting on black people in the forties or whenever they did. All of that stuff carries over...I think it is still happening today .... they need you as a guinea pig. They know you're going to die anyway. They are not trying to get you well (Jacobs et al., 2006, p. 94).

6.3 Suffering May Have Meaning

Particularly, in the African-American community, there is a reluctance to use hospice and palliative care. Some of this hesitation likely stems from a different view of pain and physical distress at the end-of-life compared with the dominant White European community. While many White Europeans may view suffering as a state of physical distress over which the individual has no control, members of other cultures have found meaning in suffering. The fear of being in pain, without being able to control it, seems to be a frequent theme among those requesting physician expedited death as well as those patients who develop advance directives. Several studies within the African-American community suggest that many African Americans may view suffering from a spiritual perspective. Rather than being "pointless," physical suffering may be meaningful and redemptive (Kagawa-Singer & Blackhall, 2001).

In the Christian tradition, some patients may view pain as a test of their faith and as sharing in Christ's suffering on the cross (Black & Rubinstein, 2004). Among

Black Caribbean interviewees with cancer (Koffman, Morgan, Edmonds, Speck, & Higginson, 2008), this theme was present with one woman analogizing her pain from breast cancer to the tribulations of Job in the Old Testament. "In some way, I think he, he's tested me…To see how strong I am, how strong my faith is, how much I believe in him…" (p. 355). By transitioning from aggressive to palliative care or establishing an advance directive to limit treatment, African-American patients may view themselves as abandoning their relationship with God and choosing an action that demonstrates a loss of faith (Phelps et al., 2009).

6.3.1 The Cultural Impact of the Holocaust on End-of-Life Decision-Making in Israel

Within the Jewish community, personal suffering may also have religious significance. Sturman (2003) observed in Israel that Jewish patients' and physicians' decisions at the end of life were influenced by the Holocaust experience. In one case, an 89-year-old rabbi, was on a respirator and "clinging to life." His family, who included a number of devout members, believed that his tenacity stemmed from his guilt from surviving the Holocaust while many of his family did not. It was suggested that he struggled to remain alive because of his guilt and fear of having to see these family members in the afterlife.

The Holocaust also influenced Israeli physicians in their choice to aggressively treat patients who were at the end of their lives even when reversal of the patient's condition was highly improbable. The guilt and accompanying sense of helplessness from having relatives perish in Nazi concentration camps seemed to motivate providers to do everything possible to maintain patients' lives—even in the face of futility (Sturman, 2003). Additionally, a Jewish physician treating a Jewish patient elicits a shared sense of mission. Others' difficulties are not simply to be idly observed. As a community, Jewish people have a duty to one another. This duty to a fellow Jew exists even if the patient does not want treatment; the obligation to do everything possible to save their life remains (Sturman, 2003).

The sacredness and value placed on life also reflected an often, unspoken, agenda of replacing Jewish lives lost in the Second World War: "There is a reluctance to give up on any patient, no matter how hopeless the case may be, since every life…helps to make up for those lost in the Holocaust" (Sturman, 2003, p. 98).

6.4 Life Is in God's Hands

Patients with a strong spiritual background also emphasized that God's power was far greater than that of the physician. There was a belief that miracles could occur and that healing may occur through God's hand regardless of medical intervention.

In a sample of surrogate decision-makers and patients, nearly 2/3 reported that God influenced the outcome of their loved one's hospital course. A common theme was that physicians could not accurately prognosticate because a loved one's outcome was predetermined by God: "I think it's whether God says it's [the patient's] turn to die. If you feel that God's in control, and I do, then no matter what a doctor will tell me, or what a doctor says, he's only human. He doesn't have all the answers. And I believe that God does" (Zier et al., 2008, p. 4). Strong faith in God is seen as overriding the physician's prognosis: "All I can do is pray for her to continue to get better and maybe one of these days, she can walk out of here." "I'm very optimistic because I know our faith is strong." (Ernecoff et al., 2015; p. 1665).

While predetermination themes were common, other surrogates believed that God could directly intervene in the patient's hospital course and that physicians' judgments were of limited validity: "…[Physicians] can say, well this person is not going to survive… and then, here comes God plays a role and just pick him up. Could be on their dying bed getting you know, CPR or anything and they think they gonna lose 'em, flatline. And they just jump back, with a heartbeat, I think that's the hands of God (Zier et al., 2008, p. 5).

In the U.S., physicians, themselves, varied considerably in their responses to family members' statements of faith in God. While patients and their families often viewed their circumstances in religious or spiritual terms, health care professionals were often not responsive to this perspective. Of 40 family meetings in an intensive care unit, 15 of them included religious content from the patient's surrogates. When family members raised spiritual or religious issues, the physician often responded with biomedical content as in the following exchange. After a surrogate indicated that she was praying for the patient not to require a tracheostomy, the physician responded "the long-term question is how to prevent the pancreatitis from happening again. It's not a question for now but it's gonna be a question pretty soon, I think." (Ernecoff, Curlin, Buddadhumaruk, & White, 2015, p. E5).

Some physicians responded more sensitively and respectfully. After a surrogate indicated that God is most powerful, the physician responded with "I agree. He is more powerful." (Ernecoff et al., 2015, p. E5). However, in general, when surrogates raised religious or spiritual issues, physicians typically did not explore or speak further about the topic. A common physician response was to ask the family if they would like to speak with a hospital chaplain. Occasionally, a physician would ask if spirituality was important to a particular patient and encouraged the family to invite any clergy they thought would be helpful (Ernecoff et al., 2015, p. E5).

6.5 Family Loyalty

6.5.1 Protecting a Seriously Ill Loved One from Emotional Harm

While perhaps viewed as deceptive to Western clinicians and ethicists, withholding information from seriously ill patients is almost always guided by compassion. Zhao (2015) provides a very touching story of protecting a spouse from knowledge of their terminal condition in China. A man who had been married for 17 years with a young son was diagnosed with serious lung cancer. He underwent treatment but was not responding. His wife was worried about the emotional impact on her husband of seeing negative test results that indicated that he was not improving. In response, she actually took the laboratory reports, changed them, and copied them. This process of changing test results occurred over a period of 15 months and even included keeping a set of these altered records at the doctor's office. The story appeared in the newspaper and on the Internet under the title of "15 Months Weaving a Life of Love." The wife's actions were praised by those who heard the story. She was portrayed as both loving and strong for taking on all of the emotional distress associated with her husband's illness and protecting him from it. As he got worse, the wife did stop the deception, but the husband never commented on it. Zhao (2015) interprets the husband's response is a way of conveying his appreciation to his family and hopefully reducing their distress about his impending death (Zhao, 2015).

Protecting a seriously ill loved one or patient from emotional distress has been one of the utilitarian arguments offered for withholding information from the patient by physicians and family members. If one is terminally ill, often in pain, with impending death, does the emotional distress associated with knowing one's diagnosis and prognosis improve the quality of one's remaining life? There is evidence that patients probably do know that their illness is serious and prognosis grim. However, is there any advantage, given that life is limited, in being directly confronted with this information?

Our study of Bosnian immigrants in the United States suggested that in their home country, direct truth telling by physicians was not the norm. Additionally, there was recognition that there could be harm and certainly little value, in providing the patient with this emotionally painful news. In comparing their knowledge of the U.S. system of disclosure with physician practices in Bosnia, one respondent said:

"In this country (the U.S.), the doctor says face-to-face, you have cancer …our doctors (in Bosnia) go around (it). You might have this, you might have that. Here they cut you in the middle. They tell you, "You are sick and are going to die." (Interviewer: which way do you prefer?) Going around… Not straightforward (Searight & Gafford, 2005, p. 200). Other respondents emphasized what they perceived as needless emotional distress created for the patient by requiring them to have this knowledge and to make treatment decisions: "I think it's cruel to tell somebody you have cancer and you will die." (p. 199) "It is cruel [to ask the patient to decide]… Because the patient is in a lot of pain." (Seright & Gaffrod, 2005, p. 200).

6.5.2 Respect and Do Not Burden Elders

Particularly in Asia, family-based decision-making is seen as a way of showing both love and respect for an older family member. Filial piety demonstrates respect by removing the burdens of care from the loved one and treating them with the respect and affection that they have earned by contributing to the family's well-being for many years. The care that parents provided their children while growing up is now reciprocated. This includes doing everything possible to keep a parent alive. If adult children were perceived as remiss in this duty, there would be considerable community disapproval. Even taking on a significant financial burden may be necessary to fulfill this duty. One Korean study found that 80% of out-of-pocket healthcare fees were covered by children of patients with 8% from the patient themselves (Lee, 2015). Adult children in Korea may lose their homes and savings to pay for parents' hospital care even when parents overtly indicate that they do not want continued aggressive care.

6.5.3 Maintain the Patient's Hope

There is both implicit and explicit recognition on the part of the general public that one's hopefulness can play a role in illness recovery. Providing the patient with an explicit diagnosis may dash these hopes of recovery. This theme has emerged across multiple cultures. In Pakistan, a physician describes how she had a family conference without the patient present to describe the father's widespread metastatic cancer (Moazam, 2000). The family asked the physician not to inform the father because "how long he lives is in the hands of God in any case, and it is not right to make my father lose hope while he is so ill." (Moazam, 2000, p. 280).

Hope was also described as an important dimension among recent Bosnian immigrants to the United States. Maintaining a patient's hope was an important part of how physicians in Bosnia communicated difficult diagnoses: "[in Bosnia] the doctor is always reserved… Trying not to tell the patient. The doctors always say there's hope, there's different techniques… Hope was always given to the patient" (Searight & Gafford, 2005, p. 199).

In a study of patients on kidney dialysis—predominantly White European Canadians—hope was a common theme. At times, the patient's hope was likely unrealistic—"I don't want to live on dialysis. I want to have a normal life. Hope for me now is to get back to what I've lost… To work again, to be independent and live alone, and have the life that I had." (Davison & Simpson, 2006). The authors described a tension between patients' expectations that their physician would raise the topic of advance care planning and the patient's own focus on their current daily lives. While the patients wanted information from the physician, they were looking for information that would give them hope. Patients tended however, to avoid pressing the physician for prognostic information but would use the physician's cues to guide whether it was appropriate to raise the topic of their future illness course (Davison & Simpson, 2006).

6.5.4 Familismo

As has been clear throughout this book, lifelong devotion and meaningful engagement with one's family is a characteristic of multiple cultures including Hispanic and Asian societies. Sabogal and colleagues (1987) developed a scale to assess this construct which includes items reflecting one's responsibility to provide support to family members, trust that family members are reliable sources of support and can be counted on to help solve problems, and the extent to which individual motivation is guided by a duty to the family. The dimension of duty is reflected in scale items such as" Much of what a son or daughter does should be done to please the parents" (Sabogal et al., 1987). Of interest, when comparing two samples—one from Spain and a U.S. Hispanic sample—of adults caring for a relative with dementia, the U.S. sample's pattern of responses suggested that family burden was not associated with increased caregiver distress while in Spain, greater family burden was associated with greater symptoms of depression (Losada et al., 2006). Any significant events including life threatening illness experienced by a family member initiates a process of family members coming together to provide support and in many instances, to make decisions on the patient's behalf. Sometimes to the frustration of healthcare personnel, surgery for a member of a Hispanic family may not proceed until the family is convened. This kinship network may include godparents, family friends, neighbors and relatively distant relatives (Smith, Sudore, & Pérez-Stable, 2009).

6.5.5 Filial Piety

In Asian families, the care provided to aging parents by their adult children is a way of reciprocating for the sacrifices that have been made for them. When a close family member is ill, the family shares in the suffering. Additionally, while their loved one is undergoing treatment, the family helps moderate distress and provides familiarity as the patient deals with physicians and the often frightening hospital environment The family, in their interactions with the healthcare team, represents the patient's identity. As part of their duty to a parent, adult children also take on the responsibility of helping their loved one live a meaningful life during their final months (Wang, 2015). This is not a contractual style obligation but a deeply embedded web of mutual care. Since the family is so highly valued, giving of oneself to an aging family member is natural since one's identity is their family: "… Only when the individual abandons his/her natural personality to live in accordance with the ethical entity of a family can he/she obtain true freedom and ontological self-consciousness in the family" (Cai, 2015, p. 191).

6.6 Conclusion

Blackhall et al.'s (1995) study provided quantitative information indicating that there were differences between ethnic groups regarding the acceptability of directly informing a loved one of a cancer diagnosis as well as the desired locus of decision-making at the end of a patient's life. This pattern of greater collectivism on the part of Asian and Hispanic families has generally held up in subsequent studies. However, numerical percentages do not describe the underlying reasons for these patterns. Additionally, in the case of African-Americans in the United States, the desire for aggressive treatment, rejection of hospice and palliative care, and low rate of completion of advance directives should be understood in the context of an extended history of maltreatment and discrimination by the healthcare establishment. Understanding the reasons for these differences also should lead to more culturally sensitive patterns of communication with patients and their families at the end-of-life.

References

Black, H. K., & Rubinstein, R. L. (2004). Themes of suffering in later life. *The Journals of Gerontology Series B: Psychological Sciences and Social Sciences, 59*(1), S17–S24.

Blackhall, L. J., Murphy, S. T., Frank, G., Michel, V., & Azen, S. (1995). Ethnicity and attitudes toward patient autonomy. *JAMA, 274*(10), 820–825.

Boag, W., Suresh, H., Celi, L. A., Szolovits, P., & Ghassemi, M. (2018). Modeling mistrust in end-of-life care. arXiv:1807.00124.

Cai, Y. (2015). On family informed consent: On the legislation of organ donation in China. In R. Fan (Ed.), Family-oriented informed consent: East Asian and American perspectives (pp. 187–200). New York: Springer.

Carrese, J. A., & Rhodes, L. A. (1995). Western bioethics on the Navajo reservation: Benefit or harm? *JAMA*, 826–829.

Cheng, S. Y., Suh, S. Y., Morita, T., Oyama, Y., Chiu, T. Y., Koh, S. J., … Tsuneto, S. (2015). A cross-cultural study on behaviors when death is approaching in East Asian countries: What are the physician-perceived common beliefs and practices? *Medicine, 94*(39), 1–5.

Davison, S. N., & Simpson, C. (2006). Hope and advance care planning in patients with end stage renal disease: qualitative interview study. *BMJ, 333*(7574), 886.

Ellerby, J. H., McKenzie, J., McKay, S., Gariépy, G. J., & Kaufert, J. M. (2000). Bioethics for clinicians: 18. Aboriginal cultures. *Canadian Medical Association Journal, 163*(7), 845–850.

Ernecoff, N. C., Curlin, F. A., Buddadhumaruk, P., & White, D. B. (2015). Health care professionals' responses to religious or spiritual statements by surrogate decision makers during goals-of-care discussions. *JAMA Internal Medicine, 175*(10), 1662–1669.

Halperin, E. C. (2007). The poor, the black, and the marginalized as the source of cadavers in United States anatomical education. *Clinical Anatomy: The Official Journal of the American Association of Clinical Anatomists and the British Association of Clinical Anatomists, 20*(5), 489–495.

Jacobs, E. A., Rolle, I., Ferrans, C. E., Whitaker, E. E., & Warnecke, R. B. (2006). Understanding African Americans' views of the trustworthiness of physicians. *Journal of General Internal Medicine, 21*(6), 642.

Kagawa-Singer, M., & Blackhall, L. J. (2001). Negotiating cross-cultural issues at the end of life: You got to go where he lives. *JAMA, 286*(23), 2993–3001.

Koffman, J., Morgan, M., Edmonds, P., Speck, P., & Higginson, I. J. (2008). "I know he controls cancer": The meanings of religion among Black Caribbean and White British patients with advanced cancer. *Social Science & Medicine, 67*(5), 780–789.

Kressin, N. R., & Petersen, L. A. (2001). Racial differences in the use of invasive cardiovascular procedures: review of the literature and prescription for future research. *Annals of Internal Medicine, 135*(5), 352–366.

Lavallée, L. F. (2009). Practical application of an Indigenous research framework and two qualitative Indigenous research methods: Sharing circles and Anishnaabe symbol-based reflection. *International Journal of Qualitative Methods, 8*(1), 21–40.

Lavery, L. A., Ashry, H. R., Van Houtum, W., Pugh, J. A., Harkless, L. B., & Basu, S. (1996). Variation in the incidence and proportion of diabetes-related amputations in minorities. *Diabetes Care, 19*(1), 48–52.

Lee, S. C. (2015). Intimacy and family consent: A Confucian ideal. *Journal of Medicine and Philosophy, 40*(4), 418–436.

Losada, A., Robinson Shurgot, G., Knight, B. G., Marquez, M., Montorio, I., Izal, M., & Ruiz, M. A. (2006). Cross-cultural study comparing the association of familism with burden and depressive symptoms in two samples of Hispanic dementia caregivers. *Aging & Mental Health, 10*(1), 69–76.

Moazam, F. (2000). Families, patients, and physicians in medical decision making: A Pakistani perspective. *Hastings Center Report, 30*(6), 28–37.

Noonan, A. S., Velasco-Mondragon, H. E., & Wagner, F. A. (2016). Improving the health of African Americans in the USA: An overdue opportunity for social justice. *Public Health Reviews, 37*(1), 12.

Petros, P., Abendstein, B. & Browning, A. (2018). *International Urogynecology Journal, 29*(11), 1563–1564

Phelps, A. C., Maciejewski, P. K., Nilsson, M., Balboni, T. A., Wright, A. A., Paulk, M. E., … Prigerson, H. G. (2009). Religious coping and use of intensive life-prolonging care near death in patients with advanced cancer. *JAMA, 301*(11), 1140–1147.

Reverby, S. M. (2009). *Examining Tuskegee: The infamous syphilis study and its legacy*. Durham: University of North Carolina Press.

Rhodes, R. L., Batchelor, K., Lee, S. C., & Halm, E. A. (2015). Barriers to end-of-life care for African Americans from the providers' perspective: Opportunity for intervention development. *American Journal of Hospice and Palliative Medicine, 32*(2), 137–143.

Sabogal, F., Marín, G., Otero-Sabogal, R., Marín, B. V., & Perez-Stable, E. J. (1987). Hispanic familism and acculturation: What changes and what doesn't?. *Hispanic Journal of Behavioral Sciences, 9*(4), 397–412.

Schneider, E. C., Zaslavsky, A. M., & Epstein, A. M. (2002). Racial disparities in the quality of care for enrollees in Medicare managed care. *JAMA, 287*(10), 1288–1294.

Schulman, K. A., Berlin, J. A., Harless, W., Kerner, J. F., Sistrunk, S., Gersh, B. J., … Eisenberg, J. M. (1999). The effect of race and sex on physicians' recommendations for cardiac catheterization. *New England Journal of Medicine, 340*(8), 618–626.

Searight, H. R. (2019). *Health and behavior: A multidisciplinary perspective*. Lanham, MD: Rowman & Littlefield.

Searight, H. R., & Gafford, J. (2005). "It's like playing with your destiny": Bosnian immigrants' views of advance directives and end-of-life decision-making. *Journal of Immigrant Health, 7*(3), 195–203.

Smith, A. K., Sudore, R. L., & Pérez-Stable, E. J. (2009). Palliative care for Latino patients and their families: Whenever we prayed, she wept. *JAMA, 301*(10), 1047–1057.

Sturman, R. L. (2003). *Six lives in Jerusalem: End-of-life decisions in Jerusalem: Cultural, medical, ethical and legal considerations*. New York: Springer.

Wang, J. (2015). Family and autonomy: Towards shared medical decision-making in light of Confucianism. In R. Fan (Ed.), *Family-oriented informed consent: East Asian and American perspectives* (pp. 63–80). New York: Springer.

Washington, H. A. (2006). *Medical apartheid: The dark history of medical experimentation on Black Americans from colonial times to the present*. New York: Doubleday Books.

Zhao, W. (2015). A Confucian worldview and family-based informed consent: A case of concealing illness from the patient in China. In R. Fan (Ed.), *Family-oriented informed consent: East Asian and American perspectives* (pp. 231–243). New York: Springer.

Zier, L. S., Burack, J. H., Micco, G., Chipman, A. K., Frank, J. A., Luce, J. M., & White, D. B. (2008). Doubt and belief in physicians' ability to prognosticate during critical illness: the perspective of surrogate decision makers. *Critical care medicine, 36*(8), 2341. https://www.ncbi.nlm.nih.gov/pmc/articles/PMC2628287/pdf/nihms-85681.pdf

Chapter 7
Conclusion: Continuing Changes in Marriage and Family; Supporting Diverse Perspectives on End-of-Life Decision-Making

As is evident in the case of Mrs. Kim from Chap. 1, when confronted with serious illness, family relationships are a key element of patient decision-making. While clinicians, including the author, have encountered situations where the family and the patient are at odds with one another about optimal treatment, it is important that healthcare providers not necessarily assume that active family involvement is a type of corrupting influence on individual patient autonomy. In the now classic report of the *President's Commission for the Study of Ethical Problems in Medicine and Biomedical and Behavioral Research*, the Commission did conclude that if the patient freely and knowingly gives decision-making authority to another person such as a family member, it is permissible to exclude the patient. However, as I have written elsewhere (Searight & Gafford, 2005; Searight, 1992), it is important that this preference for family based decision-making be clearly documented in the patient's record. Additionally, the healthcare provider should clearly indicate to the patient that if they change their mind and would like to be directly informed of their health status and treatment options, the physician will immediately do so. It is important to recognize that these patients have not surrendered their autonomy nor sacrificed their independent decisional authority by delegating healthcare communication to their family. These patients have made a decision not to be informed.

While certainly not of life or death magnitude, many of us, at times, particularly when feeling overwhelmed, delegate decisions to others. Consider going out to dinner with family after a very demanding day at work. You look over the menu and listen to the waiter describe the day's specials but are not really paying attention and at some level, your choices for the next meal are a blur. Having to make one more decision that day, even if it is between the fried chicken and prime rib special feels like an overwhelming chore. With an exhausted facial expression, you turn to your spouse or partner and say, "You know what I like; please order for me; I cannot make another decision today." In this common scenario, you have not sacrificed your autonomy. Instead, as a means of eliminating another cognitive and emotional burden, you have made an informed decision to delegate your dinner choice to a competent adult that you trust.

H. R. Searight, *Ethical Challenges in Multi-Cultural Patient Care*, SpringerBriefs in Ethics, https://doi.org/10.1007/978-3-030-23544-4_7

7.1 The Changing Family and Informed Consent

In many situations, it would be relatively straightforward to have the family involved as a healthcare decision-maker. For example, as noted earlier, in organ donation decisions in Taiwan, a recommended practice is to have both the patient and potential donor as well as a family member sign the donor authorization form.

However, as family relationships in the West have become more tenuous and unstable, requiring a family members' involvement in key healthcare decisions may, in some situations, not be in the patient's best interest. Family involvement in decisions is predicated on the value that their decisions are genuinely guided by what they see as the patient's best interests. Macklin (1987) provides a number of examples of situations involving substituted judgment in which the decision-maker's self-interest appears to take priority over the well-being of the patient. For example, marriages may end and partners go on to new relationships without formal divorce or legal separation. In the absence of a clear record denoting a decision-maker, in many states, if the issue is legally pressed, a spouse, even though estranged from the patient for many years, would legally be the surrogate decision-maker for the patient. While one would hope that a surrogate's end-of-life decisions would not be guided by economic factors, it is certainly possible that the proxy being named or excluded from the patient's life insurance policy, could influence decisions made about an estranged spouse's care.

Indeed, until recently in the United States when same-sex marriages were legally recognized, the long-term partner of a seriously ill patient in a same sex relationship, had no legal standing unless they were formally named as durable power of attorney. During the AIDS crisis, situations arose in which an adult child's parent, even though cut off from their adult child for many years, was the legal decision-maker for patients with cognitive impairment from AIDS-related dementia. The estranged parent's choices overrode that of the patient's partner of 20 years.

One of the often unrecognized outcomes of greater longevity is the opportunity to have a greater number of intimate relationships in later life. For example, former President Clinton's mother was married four times. She reportedly included the surnames of all spouses in her "chosen" name: Virginia Dell Blythe Clinton Dwire Kelly. At the time of her death, she had been married to Richard Kelly for 12 years. President Clinton, who was in his late 40s at the time of his mother's death, certainly had a much longer standing and stable relationship with his mother than any of her respective husbands. However, absent a formal durable power of attorney, if unable to make medical decisions for herself, President Clinton's mother's husband of 12 years would be the surrogate decision-maker in many states.

While there is no evidence that there were issues of this type in President Clinton's family, other relationships have raised questions. The late media personality, Anna Nicole Smith, at the age of 26 married J. Howard Marshall, an 89-year-old billionaire. Thirteen months after they were married, Marshall died. Smith indicated she had been promised half of Marshall's estate and her right to pursue her interest in the estate was affirmed by the US Supreme Court. Smith died of a prescription drug overdose before the legal case was completely settled. This example illustrates how the legal definition of family relationships takes precedence over historical or emotionally-based relationships in end-of-life care. The legal system is individually focused and relies heavily on contractual documents. As the Schaivo case shows, end-of-life decisions that take place through the court system will be long, drawn out, and often elicit conflict between family members—a process unlikely to be in the patient's best interests.

7.2   The Future of End-of-Life Decision-Making

There is growing evidence that the Western emphasis on individualism and autonomy is gradually becoming more widespread internationally. This was evident in Cheng et al.'s (2015) study comparing Korea, Taiwan and Japan. The Japanese stood apart from their Asian neighbors in placing somewhat more emphasis on the individual and encouraging a norm of disclosing medical bad news to the patient themselves. Moreover, while the family's presence, particularly when physicians were providing prognostic information, was seen as very important in Korea and Taiwan, only about half of Japanese physicians viewed the family's presence as important (Cheng et al., 2015).

Internationally, while research suggests there is a growing trend towards individualism with diminishing emphasis on collectivism, communitarians, community-oriented utilitarians, and those concerned with preserving cultural diversity continue to add their voices to the discussion. Autonomy, while typically seen as empowering, and the basis of self-determination, can alternatively be construed as isolating patients from important sources of social support. Hardwig (1990) emphasizes that despite a legal system focused on assigning responsibility to individuals, seemingly personal decisions do influence others' lives. Empowering individuals cannot ignore the impact that decisions such as initiating or ceasing cigarette use, undergoing a vasectomy, or allowing hypertension to remain untreated, have on one's family and other close relationships (Hardwig, 1990, 2000).

Even more true today than when Hardwig published his original article, in a contemporary medical environment concerned with cost-containment, the family has been shouldering the increased burden of caring for aging parents and grandparents. Other trends placing greater demands on the patient's social network include progressively briefer hospitalizations with patients left to recover at home, more surgical procedures are performed on an outpatient basis, and physical therapy conducted at home or through self-directed exercise protocols. While many patients prefer being at home and home health care is far less expensive than a hospital stay, this domestic care model places greater responsibility on family members for day-to-day patient care (Hardwig, 1990).

In terms of disclosing serious diagnoses such as cancer, it is important that physicians carefully consider the context rather than automatically providing full diagnostic and prognostic information or, alternatively, withholding that information from the patient. Shahdi (2010) notes that assumptions about the preferences of patients should not be made based upon ethnicity, gender or age. A systematic review of the question of physician disclosure of serious illness found that there was greater diversity in preferences among non-White European samples than among White Europeans (Larkin & Searight, 2014). Thus, any choice to allow the patient's culture to dictate how to proceed with end-of-life discussions may well be incorrect. For physicians, despite greater attention to the topic in medical schools, residencies and continuing education programs, disclosing a terminal diagnosis to a patient and/or their family, still remains, a difficult, emotionally charged, challenge. It is understandable that this anxiety might be greater with patients of ethnic backgrounds different from that of the physician, themselves. Nondisclosure will hopefully be guided by patient preference rather than by unrecognized physician anxiety about an emotionally difficult patient and family encounter.

While collectivist cultures are moving towards individualism, recent writing on physician disclosure of serious illness in the United States has suggested that the medical community, influenced by a utilitarian ethic, is re-examining the value of limited sharing of information with patients (Epstein, Korones, & Quill, 2010). If a physician's disclosure has the potential only to harm and the content is deemed by the physician to have little meaningful value for patient decision-making, it may be ethically permissible to withhold the information. (Epstein, Korones, & Quill, 2010). In recent years, there have been thoughtful commentaries by U.S physicians, including those who have developed protocols for disclosing medical "bad news," about withholding medical information from patients when appropriate. Information which may not be beneficial to the patients include the potential for low probability adverse events associated with a procedure that can obscure key decisions that the patient needs to make. In circumstances in which information is not disclosed, it is necessary to reflect upon the balance of beneficence and nonmaleficence. In these discussions, when a patient is silent or changes the subject, it likely that they are communicating a desire not to be provided with additional information at this time (Epstein, Korones, & Quill, 2010). Probably one of the most common examples of this phenomenon are pharmaceutical advertisements on television in which there is a rapid recitation of all possible side effects—often including death. Another situation in which information may be withheld from is when non-urgent incidental findings arise while the patient is being investigated for a condition of significance such as cancer.

Epstein et al. (2010) also make the important point that information "dumps," in which the legal requirements of reciting all possible risks take precedence over patient understanding, may not necessarily enhance patient autonomy. In fact, because of the inability to prioritize large amounts of unfamiliar, yet personally relevant, medical information, providing all details to follow a guideline of complete disclosure may make it difficult for the patient to participate meaningfully in their care.

In the United Kingdom, the General Medical Council's (2008) policy is that, in general, patients should be aware of their medical condition, their prognosis and treatment options. However, in situations involving dialysis, the General Medical Council supports physicians if they determine that treatment would be non-beneficial.

Additionally, if a judgment is made that the treatment would not be beneficial, physicians are not required to disclose options such as dialysis and actually may withhold information about potential treatments if they believe that providing the information might psychologically harm the patient. For example, if a patient is in need of stent placement but has been acutely anxious about the possibility of having a heart attack like the one afflicting their younger brother, the procedure actually may go ahead without complete informed consent since the patient could refuse treatment deemed medically necessary. In practice, how can a clinician balance the legal and ethical requirements for informed consent with either personal or culturally held views about non-disclosure of medical "bad news?"

7.3 Recommendations

In considering the accounts of family-based decision-making with concepts such as filial piety guiding surrogate decision-making, it is clear that these models, which continue to be common in Asian and Hispanic cultures, are inconsistent with individually focused living wills and durable powers of attorney in their current forms. In China, it has been suggested that appointing someone as a proxy decision-maker such as a durable power of attorney, should be a collective decision of the family. Importantly, while the durable power of attorney is certainly recognized by the patient' healthcare team, the person deemed DPOA should have the support of the extended family. Any written advance directive should be signed by the patient and all significant family members (Yang, 2015).

Padela and colleagues have called for a second order autonomy consistent with multicultural health care in an increasingly "flat" word (Padela, Malik, Curlin, & De Vries, 2015). In this approach, the patient would voluntarily choose to have choices made about their care by their family. They may also choose to have all information shared with family members as well (Padela et al., 2015). This choice can be documented with signatures from the patient, healthcare professionals, and relevant family members. This approach would meet the legal requirements for documented informed consent in countries such as the United States while recognizing alternative approaches to end-of-life decision-making. In countries such as Pakistan, where families appear to want the physician to make the decision (Moazam, 2000), relevant family members can document their agreement.

In discussions with patients, a useful starting point is to ask the patient directly about the amount of information they would like: "Some people would like a good deal of information, others do not; What about yourself?" In order to assess whether patients prefer family-based or even physician centered decision-making, they can be asked directly about how they would like decisions to be made. Physicians should encourage patients to indicate the person or persons that they trust in assisting them to make good quality medical decisions (Brown, Bekker, Davison, Koffman, & Schell, 2016).

In terms of ethical theory, principalism is spreading well beyond the borders of the United States and northern Europe. Principalism, when applied in its "pure" form, could guide decision-making in multiple cultural contexts. The rationale for protecting patients from a terminal diagnosis can be accounted for through beneficence and nonmaleficence. Additionally, in applying the model, clinicians and ethicists should be able to justify decisions in terms of prioritizing the principles. International and cross-cultural acceptance of principlism, however, would require that autonomy is no longer "first among equals" (Gillon, 2003) but is weighted equivalently with the other three principles.

References

Brown, E. A., Bekker, H. L., Davison, S. N., Koffman, J., & Schell, J. O. (2016). Supportive care: Communication strategies to improve cultural competence in shared decision making. *Clinical Journal of the American Society of Nephrology, 11*(10), 1902–1908.

Cheng, S. Y., Suh, S. Y., Morita, T., Oyama, Y., Chiu, T. Y., Koh, S. J., … & Tsuneto, S. (2015). A cross-cultural study on behaviors when death is approaching in east asian countries: What are the physician-perceived common beliefs and practices? *Medicine, 94*(39), 1–5.

Epstein, R. M., Korones, D. N., & Quill, T. E. (2010). Withholding information from patients—when less is more. *New England Journal of Medicine, 362*(5), 380.

General Medical Council (2008). https://www.gmc-uk.org/ethical-guidance/ethical-guidance-for-doctors/consent/part-2-making-decisions-about-investigations-and-treatment#paragraph.

Gillon, R. (2003). Ethics needs principles—four can encompass the rest—and respect for autonomy should be "first among equals". *Journal of Medical Ethics, 29*(5), 307–312.

Hardwig, J. (1990). What about the family? *Hastings Center Report, 20*(2), 5–10.

Hardwig, J. (2000). *Is there a duty to die?: And other essays in bioethics.* Routledge.

Larkin, C., & Searight, H. R. (2014). A systematic review of cultural preferences for receiving medical "bad news" in the United States. *Health, 6*(16), 2162.

Macklin, R. (1987). *Mortal choices: Ethical dilemmas in modern medicine.*

Moazam, F. (2000). Families, patients, and physicians in medical decision making: A Pakistani perspective. *Hastings Center Report, 30*(6), 28–37.

Padela, A. I., Malik, A. Y., Curlin, F., & De Vries, R. (2015). [R e] considering respect for persons in a globalizing world. *Developing World Bioethics, 15*(2), 98–106.

Searight, H. R. (1992). Assessing patient competence for medical decision making. *American Family Physician, 45*(2), 751–759.

Searight, H. R., & Gafford, J. (2005). "It's like playing with your destiny": Bosnian immigrants' views of advance directives and end-of-life decision-making. *Journal of Immigrant Health, 7*(3), 195–203.

Shahidi, J. (2010). Not telling the truth: Circumstances leading to concealment of diagnosis and prognosis from cancer patients. *European Journal of Cancer Care, 19*(5), 589–593.

Yang, Y. (2015). A family oriented confucian approach to advance directives in end-of-life decision-making for incompetent elderly patients. In R. Fan (Ed.), *Family-oriented informed consent: East Asian and American perspectives* (pp. 257–270). New York: Springer.

Bibliography

Arraras, J. I., Illarramendi, J. J., Valerdi, J. J., & Wright, S. J. (1995). Truth-telling to the patient in advanced cancer: Family information filtering and prospects for change. *Psycho-Oncology, 4*(3), 191–196.

Battin, M. P., Van der Heide, A., Ganzini, L., Van der Wal, G., & Onwuteaka-Philipsen, B. D. (2007). Legal physician-assisted dying in Oregon and the Netherlands: Evidence concerning the impact on patients in "vulnerable" groups. *Journal of Medical Ethics, 33*(10), 591–597.

Blendon, R. J., Hyams, T. S., & Benson, J. M. (1993). Bridging the gap between expert and public views on health care reform. *JAMA, 269*(19), 2573–2578.

Bowman, K. W., & Hui, E. C. (2000). Bioethics for clinicians: 20. Chinese bioethics. *Canadian Medical Association Journal, 163*(11), 1481–1485.

Bradley, E. H., Wetle, T., & Horwitz, S. M. (1998). The patient self determination act and advance directive completion in nursing homes. *Archives of Family Medicine, 7*(5), 417.

Finkelstein, E. A., Bilger, M., Flynn, T. N., & Malhotra, C. (2015). Preferences for end-of-life care among community-dwelling older adults and patients with advanced cancer: A discrete choice experiment. *Health Policy, 119*(11), 1482–1489.

Kaiser, K., Rauscher, G. H., Jacobs, E. A., Strenski, T. A., Ferrans, C. E., & Warnecke, R. B. (2011). The import of trust in regular providers to trust in cancer physicians among white, African American, and Hispanic breast cancer patients. *Journal of General Internal Medicine, 26*(1), 51–57.

Kapadia, F., Singh, M., Divatia, J., Vaidyanathan, P., Udwadia, F. E., Raisinghaney, S. J., … & Karnad, D. R. (2005). Limitation and withdrawal of intensive therapy at the end of life: Practices in intensive care units in Mumbai, India. *Critical Care Medicine, 33*(6), 1272–1275.

Kelton vs. Washington DC Court of Appeals (April, 1980). 413 A.2d 919 (D.C. 1980).

Ko, E., Roh, S., & Higgins, D. (2013). Do older Korean immigrants engage in end-of-life communication? *Educational Gerontology, 39*(8), 613–622.

Makino, J., Fujitani, S., Twohig, B., Krasnica, S., & Oropello, J. (2014). End-of-life considerations in the ICU in Japan: Ethical and legal perspectives. *Journal of Intensive Care, 2*(1), 9.

Searight, H. R. (2017). Clinical and ethical issues in working with a foreign language interpreter. *Journal of Health Service Psychology, Fall.* Retrieved from https://www.nationalregister.org/pub/the-national-register-report-pub/journal-of-health-service-psychology-fall-2017/clinical-and-ethical-issues-in-working-with-a-foreign-language-interpreter.

Seoane, A. (2011). Advance directives in Spain. In S. Negri (Ed.), *Self-determination, dignity and end-of-life care* (pp. 299–330). Leiden: Martius Nijhoff.

Wachterman, M. W., McCarthy, E. P., Marcantonio, E. R., & Ersek, M. (2015). Mistrust, misperceptions, and miscommunication: A qualitative study of preferences about kidney transplantation among African Americans. *Transplantation proceedings, 47*(2), 240–246.